POLYPORES OF KERALA

POLYPORES OF KERALA

K.M. LEELAVATHY
and
P.N. GANESH

2000

Daya Publishing House
Delhi-110035

ISBN 81-7035-221-5

Published by : **Daya Publishing House**
1123/74, Deva Ram Park
Tri Nagar, Delhi-110 035
Phone : 7103999, 7199029
Fax : (011) 7198902

Printed at : **Chawla Offset Printers**
Delhi-110 052

PRINTED IN INDIA

ACKNOWLEDGEMENTS

The encouragement and help received from Prof. Anjali Roy, Santhinikethan an expert on Polypores of India, throughout the period of our study is gratefully acknowledged. The authors express their deep sense of gratitude to Prof. Lief Ryvarden, Oslo, Norway, for his unstinted help in the identification or confirmation of the identity of a large number of specimen. The Generosity of Mr. Sujan Singh and Mr. Balwant Singh, Forest Pathology Division, Forest Research Institute, Dehra Dun in making available the collections of polypores in the herbarium there for comparison and also in helping in the identification of some of our collections helped immensely in the initial part of our studies. We are grateful to them both. Dr. Derek A Reid, Royal Botanic Gardens, Kew, U.K. who helped in the identification some of our collections is also gratefully remembered.

Our colleagues and friends in the Department of Botany, University of Calicut have always supported and encouraged us in our studies. Prof. M. Sivadasan, who helped in identification of host plants, Dr. P. Manimohan, who helped in photography and Mr. Agratious Thomas who helped with some of the drawings need special mention.

But for the financial assistance from DST, Govt. of India the publication of this book would not have been possible. The first author gratefully acknowledges the generous help from the DST. Prof. K.S. Manilal, Department of Botany, University of Calicut who advised and encouraged the first author to seek financial assistance from DST for the publication of this monograph is also gratefully remembered.

The University of Calicut, which permitted the use of laboratory and library facilities and Prof. K. Unnikrishnan, the Head of the Department of Botany who provided all necessary facilities is gratefully acknowledged.

The authors also express their sincere thanks to Mrs. G. Usha Devi who did the major part of typing the manuscript and also helped in the preparation of figures and plates and to Mr. Anil Mittal of M/S Daya Publishing House, Delhi for agreeing to publish the book.

Last but not least we thank the many workers on polypores from all over the world for the prompt response to our request for necessary literature.

PREFACE

When we started working on polypores of Kerala and wanted to identify our collections, we faced many difficulties, the foremost being lack of any earlier work on this group of fungi from Kerala. The only book available was polypores of India by Bakshi (1971). Except for a very recent book by Roy and De (1996), Bakshi's book still remain the major source of information on the subject. A persual of these books brought out the fact that the fungi described in these are mostly confined to collections from the northern parts of India. Out of a large number of specimens studied by Bakshi (1971) and Roy and De (1996) only about 45 species of polypores have been recorded from whole of South India and out of these the reports from Kerala is a meagre six species of polypores. In a recent compilation (Gowda et al., 1996) of fungi reported from Kerala the number of polypores recorded remains the same six. It was this paucity of information on this important group of fungi from Kerala that challenged the present authors to take up the ambitious task of presenting the results of our study of the polypores of Kerala in the form of a book. As the period of our study and the area covered are limited, this work is by no means the last word on polypores of Kerala. The fact that even this limited study could record some genera and species new to India gives ample scope for a more comprehensive study of the polypore flora of these regions. If it prompts some young mycologists to study more of the polypores of this floristically rich and distinct region of India our purpose in presenting this book will be served.

The anxiety caused by the publication of a book on polyporaceae of India as recently as 1996 was belied when we found that a number of genera and species recorded from Kerala are new reports for India. Therefore, besides being a regional flora, it adds considerably to our knowledge of polypores of India.

CONTENTS

INTRODUCTION

Fungi of various types normally inhabit the different parts of a plant body such as roots, stem and leaves. Some of these may be harmless saprophytes while others may be weak or dangerous pathogens. A good number of these fungi produce large and conspicuous fruitbodies and are therefore called as macrofungi. These macrofungi mainly belong to two orders, Aphyllophorales and Agaricales, of the class Basidiomycetes while a few are Ascomycetes. Many of these macrofungi cause plant diseases and wood decay which are responsible for great economic loss in forestry as well as destruction of stored logs and wood products in common use.

The polypores are Basidiomycetes producing holobasidia and ballistosporic basidiospores typically on the inside of the tubes lining the underside of the fructification. Most of the polypores are wood inhabiting and the rest are terrestrial. The mycelium at first grows well, ramifying the woody tissues, absorbing nutrients after breaking the cell wall constituents by enzymatic activity resulting in decay. Most dangerous is the 'heart rot' which establishes itself in the heartwood and progresses with time. The rot causing polypores can be identified from the fruitbodies that develop on the affected trees later on. The fruitbodies are very conspicuous as brackets, mushrooms or crusts on standing or fallen wood or branches of trees on which they grow.

It is the breaking down of the wood and changes in its physical and chemical properties that is termed wood decay. From Nature's point of view the decay fungi play an important role in cyclic processes that replenish carbon-di-oxide and other substances essential for the plant growth, which in turn support animal life. They are the major agents of biodegradation which add fertility to the soil. Polyporoid species are among the main wood decay fungi encountered in nature.

There are two broad classes of wood destroying fungi, the first causing 'white rot' decomposes all components of the wood including lignin while the second causing 'brown rot' decomposes the cellulose and its associated pentoses, leaving the lignin more or less unaffected. In the first, wood is reduced to a white fibrous spongy mass while in the second it becomes a carbonous mass in varying shades of brown (Peace, 1962).

A perusal of literature reveals that due to the great economic importance of wood decaying fungi in forestry, considerable work on this group of fungi has been done in Europe, America and Australia while tropical wood rotting fungi have been comparatively less studied. Our knowledge of Indian polypores is confined to those of the northern regions of the country, mainly Western Himalayas and Bengal while very little is known of this group from the rest of the country. Therefore, study of this important group of wood inhabiting fungi of Kerala was undertaken with the assumption that Kerala State, which lies in the west coast of Southern India, floristically very rich and phytogeographically distinct from other parts of India, may have a polypore flora of it own.

It has to be emphasised here that this treatise is only a preliminary study of the polyporoid fungi occurring in Kerala and that this report is far from complete as the study was confined to collections from the central and northern parts of Kerala. Based on over five hundred collections, seventy nine species in thirty two genera belonging to three families viz. Ganodermataceae Donk, Hymenochaetaceae Donk and Polyporaceae Corda of the order Aphyllophorales are reported here. This is expected to serve as an initial step towards better understanding of the polypores of Kerala.

MATERIALS AND METHODS

Majority of the specimen described in this treatise were collected by the authors during the period 1983-1987 while a few collections were made more recently, from the forests as well as inhabited areas of central and northern Kerala which includes the political districts of Cannanore, Calicut, Wyanaad, Malappuram, Palghat, Trichur and Kottayam (Fig. 1.). The collections were made mainly in the humid months, June to November, the period when Kerala experiences rains due to South-west monsoon and North-east monsoon.

The fruitbodies were collected along with the supporting wood. General macrocharacters of the fruitbody including colour of different tissues and the type of rot were noted in the field itself. Some of the host trees were identified in the field itself while for others the help of books or experts was needed. Fruitbodies were wrapped in paper bags and brought to the laboratory. Spore deposits were taken on microslides by keeping the fresh fruitbody in humid condition. Measurements and detailed observation of the fruitbodies were made in the laboratory and the materials dried in a hot air oven at 70°C. Representative portions of the dried specimens were used for microscopic studies. The rest of the fruitbody along with the rotten wood were treated with Mercuric Chloride against mites and moulds and stored with moth-balls in paper bags as voucher specimens. The bags were kept in almarahs at room temperature. The specimens were given collection numbers with dates and details of the locality from where they were collected.

Microscopic details of various representative areas of the fruitbody, hyphal system and spores were studied as suggested by Teixeira (1962). Five per cent KOH

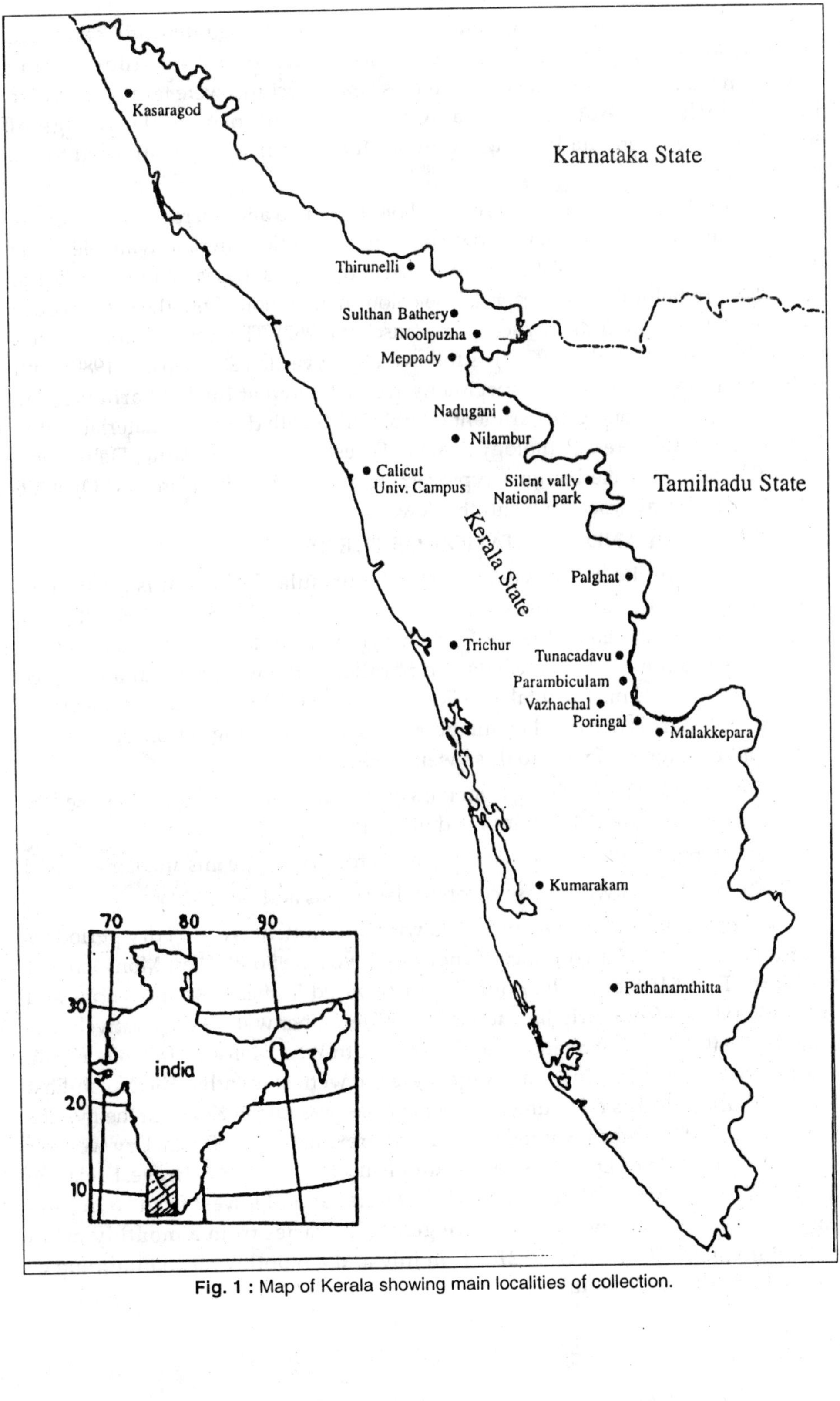

Fig. 1 : Map of Kerala showing main localities of collection.

and lactophenol were used as a general mounting media. Congo-red and Cotton blue strains were routinely used. Amyloid reaction of the spores was studied using Melzer's reagent. Measurements of all microscopic structures were taken from KOH mounts with a Meopta trinocular Compound microscope. Drawings of microstructures were made using a Camera-lucida (mirror type) attached to the microscope.

Identifications of the fungi were based on the macro and microcharacters of the fruitbody and the type of rot. Comparison with authentic materials available in the Mycology Herbarium and National type collection of Forest Research Institute, Dehra Dun was very helpful in this respect. The colour names along with the colour codes added in brackets are from Kornerup & Wanscher (1967). The classification adapted essentially is that of Pegler (1937b) as followed by Ryvarden & Johansen (1980) with minimum modifications. The specimens are deposited at the Herbarium of the Calicut University Botany Department (Herb. CALI) with duplicate materials at the Herbarium of the Forest Pathology Branch, Forest Research Institute, Dehra Dun (Herb. FRI) and at the Herbarium Cryptogamme Indae Orientalis (Herb. HCIO) of the Indian Agricultural Research Institute, New Delhi.

PHYSIOGRAPHY AND VEGETATION OF THE REGION

Kerala lies in the south-west corner of peninsular India and is positioned between 8° 18′ and 12° 4′ north latitudes and between 74° 52′ and 77° 22′ east longitudes. It is bounded by the Western Ghats mountain ranges to the east and the Arabian Sea to the west. The Ghats run parallel to the west coast at a distance ranging from 40-80 km. The total area is approximately 38,863 square kilometers of which a good part is occupied by inland waters. Altitude ranges from sea level to 5,870 m and the terrain falls into three well marked divisions :

1. the hilly tracts undulating from the Western Ghats in the east and marked by long spurs, extensive ravines and dense forests,
2. the cultivated plains intersected by numerous river, streams and lagoons and
3. the coastal belt with dense coconut plantations and rice fields.

The tropical humid climate of Kerala with intermittent wet and dry periods is closely associated with two principal monsoons, viz. the South-West Monsoon and the North-East Monsoon. The rainfall is distributed in these two monsoons and generally extends from early June to middle of December with a comparatively less rainy interval in September and October. The South-West Monsoon first hits Kerala coast during June. June, July and August are the wettest months. The North-East Monsoon, also called as retreating monsoon, brings less rain in Kerala in the months of October and December. Annual rainfall is approximately 3,000 mm. Dry weather starts from middle of December and extends up to early June. Highest relative humidity varies from 100 in June to 80 in December and lowest from 78 in July-August to 48 in January. Temperature generally varies from a monthly mean maximum of 33.6°C in April to 27.7°C in July and a monthly mean minimum of 26.5°C in April to 21°C in July.

VEGETATION

(The following account is adapted from Chandrasekharan, 1962 a,b,c,d).

The vegetation of Kerala is unusually rich and diverse, almost unique for so small a region. India is divided into several phytogeographic regions based on static features of geography and physiography. Kerala is in the Malabar region which extends along the western side of the peninsula from southern Gujarat to Kanyakumari. Due to different geological history and relative isolation, the flora of the west coast differs considerably from other region of India. Bordering the eastern side is the Deccan plateau which is considered as the 'rain shadow region' of the Western Ghats. Major parts of the other southern Indian states like Tamil Nadu, Andhra Pradesh and Karnataka belongs to the Deccan plateau.

The abundance of species of the Malayan-type, especially plants of Sterculiaceae, Teliaceae, Anacardiaceae, Meliaceae, Myrtaceae, Melastomaceae, Gesneriaceae, Piperaceae, Orchidaceae and Araceae is one of the most distinctive characteristics of Malabar flora. The presence of taxa belonging to families Guttiferae, Dipterocarpaceae, Myristicaceae and many species of Palmae and bamboo is also distinctive. Another noteworthy feature is the endemism of some species such as *Gluta travancorica* Bedd., *Dipterocarpus bourdillonii* Brand., and the monotypic genus *Blepharistemma* Wall. ex Benth. which are found only in Kerala, though allied species are present in the Andamans, Burma and Indo-Malayan regions.

Different vegetational types can be met with in Kerala depending on altitude and consequent variations of rainfall and temperature. In a cross section of the State three main vegetational types can be seen :

(a) Deciduous forests or monsoon forests from sea level to about 610 m, sometimes rising up to 915 m.

(b) Tropical evergreen forests or rain forests between 762 m and 1067 m, sometimes starting from 610 m itself and merging into the monsoon forests.

(c) Temperate evergreen forest formations at an elevation of above 1524 m, with all conceivable grades between the main zones.

Most of the natural forests are fast degrading due to deforestation and other human interference and true virgin forests are almost lacking.

HISTORICAL REVIEW OF LITERATURE ON POLYPORES OF INDIA

The taxonomic study of Indian polypores was initiated by European scientists towards the middle of the nineteenth century. The earliest report seems to be that of Klotzch (1832, 1833). A large number of polypores was reported by Berkeley (1839, 1850, 1851, 1854, 1855, 1866, 1872) based on his study of Dr. Hooker's extensive collection of macro-fungi from Sikkim-Himalayas. Other early reports of Indian polypores include those of Montagne (1842), Leveille (1844, 1846a, b), Cook (1876,

1881, 1891 a,b,c,) Massee (1901, 1906, 1908, 1910), Hennings (1900, 1901), Lloyd (1898-1925, 1904-1919, 1910b, 1912, 1915 a,b), Butler (1903, 1909, 1918), Theissen (1911), Blatter (1911) and Murrill (1924). In their compilation of fungi of India, Butler & Bisby (1931) lists 293 polyporoid species in 16 genera.

S. R. Bose, of Carmichael Medical College, Calcutta, was probably the first Indian who can be credited with starting a systematic study of polypores of India. He recorded a large number of fungi from Bengal (Bose, 1919-1928, 1921 a,b, 1922a, 1923, 1924, 1925, 1934, 1942, 1946) and Assam (Bose, 1937a). His contributions also include studies on the distribution of polypores at high altitudes (Bose, 1935), the geographical distribution and history of polypores in Bengal (Bose, 1922b), study of abnormal structures in some polypores (Bose, 1931, 1933 ,1936, 1937b,c, 1939) and culture studies of *Ganoderma lucidum* and *Hexagonia discopoda* from spore to spore (Bose, 1929, a,b). Further, from his wide experience with polypores, Bose (1944) suggested the use of certain characteristic anatomical features in addition to the characters of basidia and spores for the specific identification of these fungi.

The work on polypores of Bengal was continued by Banerjee and co-workers (Agarkar & Banerjee, 1933; Banerjee, 1947; Banerjee & Bakshi, 1954; Banerjee & Chakravarthy, 1945; Banerjee & Chatterjee, 1945 a,b,c and Banerjee & Ghosh, 1942-43). Some of their other contributions are on polypores of Sikkim-Himalayas (Banerjee, 1946; Banerjee & Ghosh, 1945) and devising a simple method for producing typical sporophores of *Polystictus sanguineus* (Banerjee & Sinha, 1955).

Some of the sporadic reports on India polypores are those of Hanif Khan (1910), Sher Singh (1924), Mitter & Tandon (1932, 1938), Uppal et al. 1935, Vahid (1938) and Roy (1953).

By the middle of the present century study of polypores causing diseases and decays in forest trees gained momentum at the Forest Research Institute, Dehra Dun. Bagchee and Bakshi studied extensively the polypores causing diseases of oaks and other economically important forest trees (Bagchee, 1950, 1953, 1954, 1957, 1958, 1961; Bagchee & Bakshi, 1950, 1951). Some of the important contributions of Bakshi and co-workers are their new report of polypores (Bakshi, 1956, 1965), studies on diseases and decay of conifers (Bakshi, 1955) diseases of *Acacia catechu* and its prevention (Bakshi, 1957a), heart rot in relation to management of *Shorea robusta* (Bakshi, 1957b), root rot of *Shorea robusta* caused by *polyporus shorea* and *P. ruidus* (Bakshi & Boyce, 1959), heart rot due to *P. palustris* (Bakshi & Balwant Singh, 1961) and studies on a disease-complex in teak (Bakshi et al. 1966). An excellent review on heart rot in trees (Bakshi & Sujan Singh, 1970) exposed the menace of wood decay fungi of which polypores play the major role. Other important contributions include the studies on *Trametes, Poria* and *Fomes* (Balwant Singh, 1961 1966 a,b; Balwant Singh & Bakshi, 1961; Puri, 1956), on the polypore flora of South

Andamans (Sujan Singh et al. 1961) and Nicobar Islands (Sehgal et al. 1961) and on the temperature relationships of Indian polypores (Sehgal et al. 1966). A monograph on Indian Polyporaceae on the trees and timber by Bakshi (1971) describes 355 species belonging to 15 genera.

Thind and co-workers from the Punjab University have reported several polypores including some new species from the Western-Himalayan and Mussoorie Hills (Thind & Chatrath, 1957, 1960; Thind et al. 1957, 1970; Thind & Dhanda, 1978, 1979 a,b, 1980; Thind & Rattan, 1971 a,b,c; Reid et al. 1959).

Significant contribution to anatomical studies of several polypores has been made by Roy (1968 a,b, 1969, 1971, 1972, 1975, 1976). More recently identification of polypores in culture has been undertaken by Bakshi et al. (1969, 1970) and Sen (1973), while Roy and co-workers have done interfertility and culture studies of several polypores (Roy, 1981 a,b; Roy & De, 1977, 1979, 1980; De, 1977, 1981; De & Roy, 1978, 1980, 1981).

The list of Southern Indian fungi compiled by Rangaswami and co-workers (1970) includes 44 polyporoid species belonging to 13 genera of which only 5 are from Kerala. More recently Natarjan & Kolandavelu (1985) described some resupinate Aphyllophorales from Tamil Nadu region and this includes the poroid members' *Inonotus polymorphus* (Rostk.) Pilat, *Phellinus allardii* (Bres.) Ryv., *P. umbrinellus* (Bres.) Ryv. and *P. purpureo-gilvus* (Petch) Ryv. The most recent work on polyporaceae of India is by Roy and De (1996) which gives a record of only six polypores from Kerala.

TECHNICAL TERMS

(a) **Fruitbody or basidiocarp :** Fruitbodies are of different shapes, size and consistency and differ in the type of attachment to the substratum or host. They may be resupinate, effused-reflexed, pileate or stipiate and may be woody, corky, leathery or cottony, hard or brittle, rarely watery. Though some polypores have variable fruitbodies, more often the type of fruit body in a species is consistent and is of taxonomic importance. The length of the fruitbody is the distance measured parallel to the substrate, the width from substrate to margin and thickness from pileus surface vertically to pore surface. Fruitbodies may be annual or perennial (Fig. 2.)

(b) **Pileus :** The upper surface of the pileus may be glabrous or covered by different types of hairs, even or uneven. In some it is shiny and smooth initially but becomes dull, incrusted or cracked with time. The color may also change with time, cuticle formation or whether. Many polypores have concentric zonations on the surface which reflect periods of increased and decreased growth. So it is important to collect fruit bodies at various stages of growth if possible.

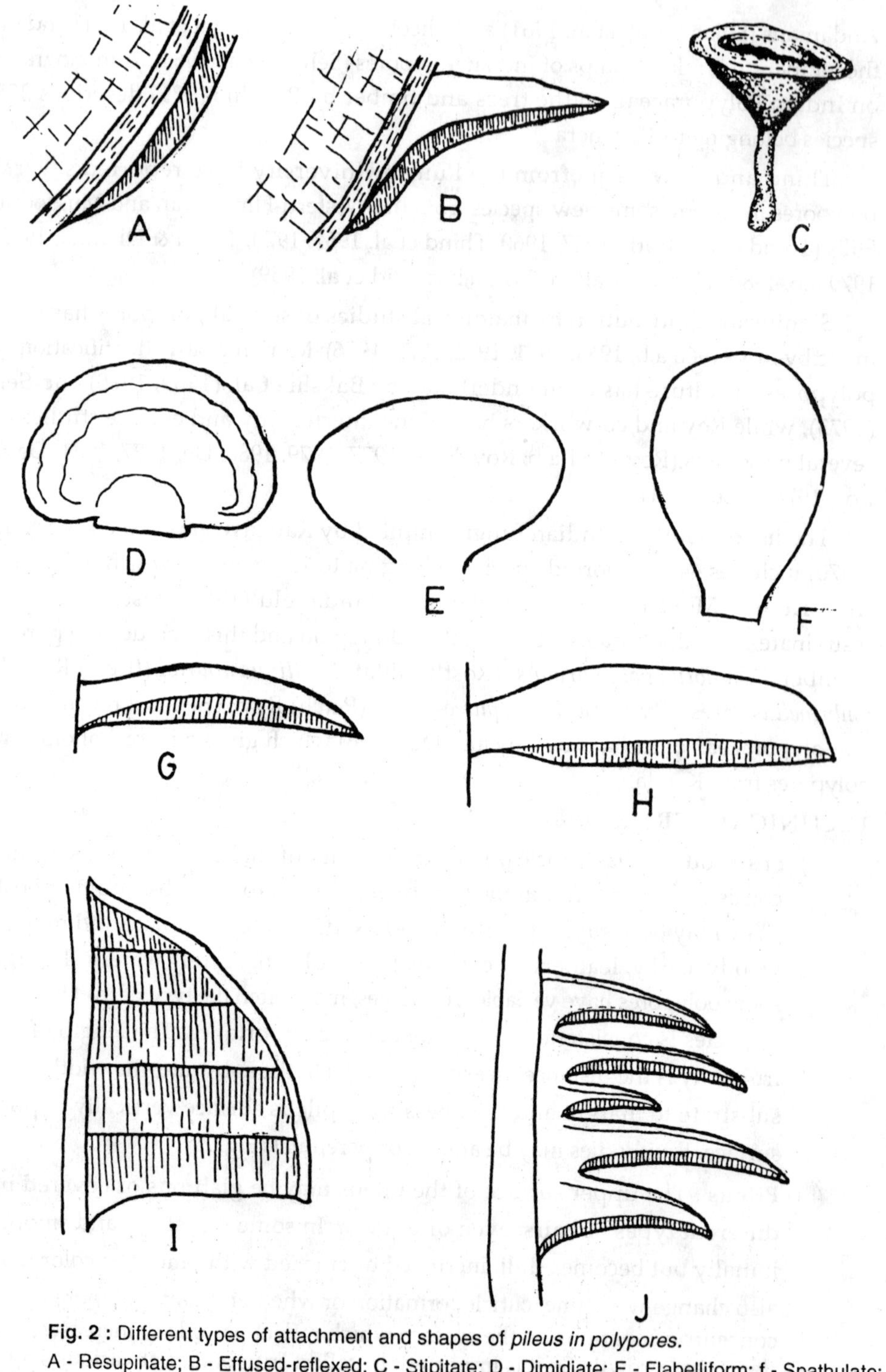

Fig. 2 : Different types of attachment and shapes of *pileus in polypores.*

A - Resupinate; B - Effused-reflexed; C - Stipitate; D - Dimidiate; E - Flabelliform; f - Spathulate; G - Conchate; H - Substipitate; I - Ungulate; J - Imbricate.

(c) **Cuticle :** Cuticle is a differentiated covering tissue consisting of a single layer of hyphae.

(d) **Crust or cortex :** It is the densely interlaced and more or less sclerotized outer layer or rind of the fruitbody, the hyphal elements of which may be modified to various extent. Often it consists of short-celled, thick-walled, coloured generative hyphae.

(e) **Pores :** The hymenophore (pore) on the lower side of the fruitbody (pileus) is very important. The color of the pore surface and the shape and size of pores differ considerably in different genera and species. Sometimes in the same fruitbody there will be transitions from one type to another, but mostly pores are consistent in their characteristic. The size of the pore is given as number per mm. Some of the common shapes of pores are given in Fig. 3.

(f) **Pore tubes :** The hymenium lines the pore tubes. The color and length of the pore tubes are important. Often the pore tubes are concolorous to pore surface and different from the color of the context. The length of the pore tube is expressed in mm.

(g) **Dissepiments :** These are the partition wall/tissue between pores, comparable to the trama in agaricales.

(h) **Context :** Context is the sterile tissue between the pore tubes and pileus surface. It may be homogeneous or duplex in consistency. In the case of duplex context, there will be a distinct difference in color and or consistency between the upper and lower parts. Whenever there is a distinct cuticle on the pileus, the context is the structure below the cuticle.

(i) **Hyphae :** The Hyphae are primarily of two types :

1. *Generative hyphae*
2. *Vegetative Hyphae*

Vegetative hyphae are of two types :

(*a*) *Skeletal hyphae*
(*b*) *Binding hyphae*

Generative hyphae is present in all species. According to the different types of hyphae present a species may be :

1. **Monomitic :** if only generative hyphae are present
2. **Dimitic :** if generative and skeletal or binding hyphae are present.
3. **Trimitic :** if generative, skeletal and binding hyphae are present.

Generative hyphae : In the mature fruitbody, it is the generative hyphae which develop a hymenium with the basidia. These are often thin-walled, sometimes thick-walled, hyaline, branched and septate. Septa may be simple or clamped. Generative hyphae easily collapse during drying and may be difficult to observe when other types of hyphae dominate in the fruitbody. Gloeopleurous hyphae are a special form of generative hyphae, usually wider than ordinary hyphae, filled with oily to resinous granular substances which often react with iodine.

Vegetative hyphae : Both skeletal hyphae and binding hyphae do not have septa.

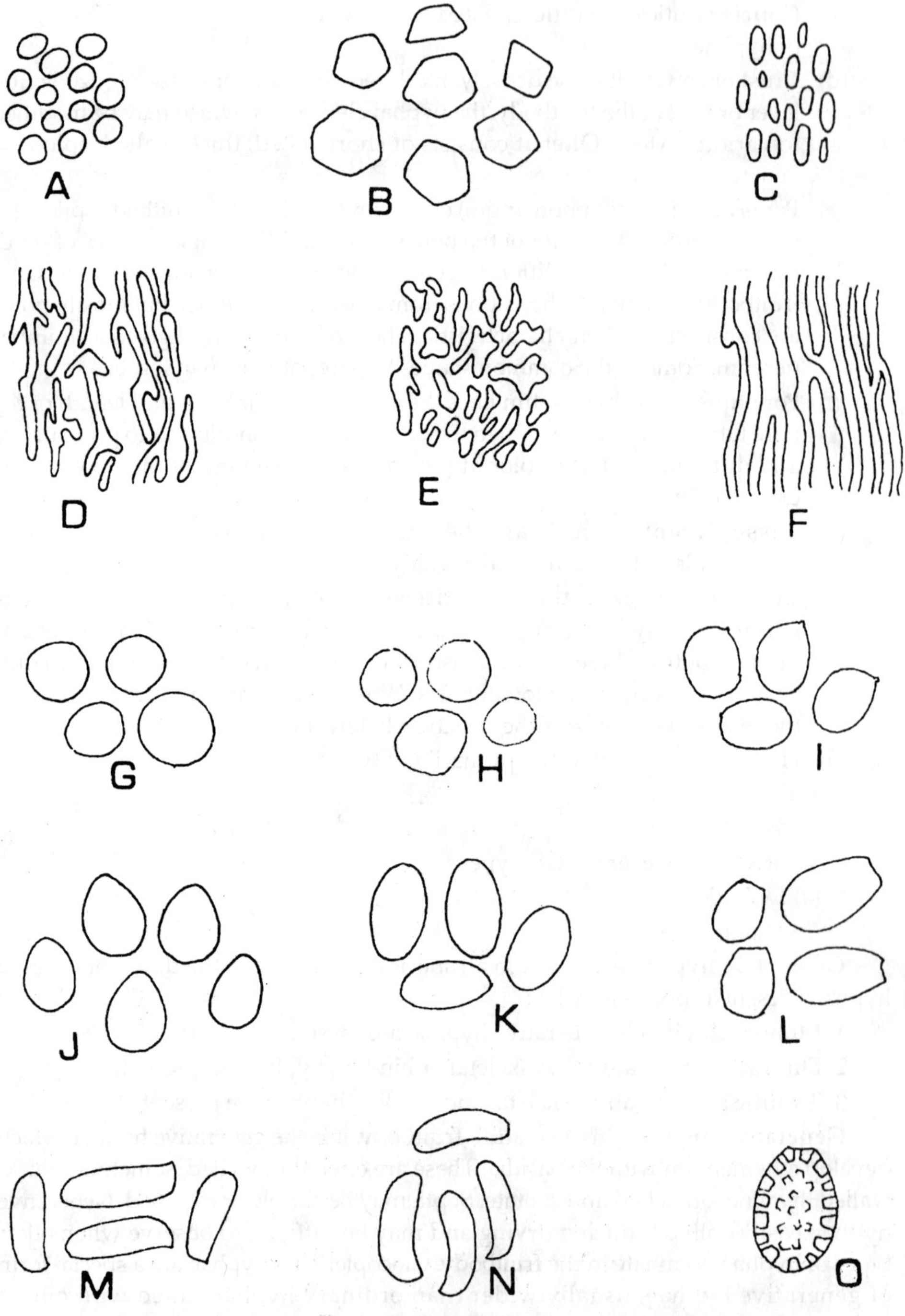

Fig. 3 : Different types of *hymenophore* (A-F) and *spores* (G-O) in *polypores.*

A - Round; B - Angular/hexogonal; C - radially elongated; D - Daedaleoid; E - Irregular; F - Lamellate; G - Sub - Globose; H - broadly ellipsoid; i - Truncate; J - Ovoid; K - Oblong ellipsoid; L - Globose; M - Cylindrical; N - Allantoid; O - Verruculose with smooth exosporium.

Skeletal hyphae : These are thick-walled to almost solid, long and unbranched, rarely with few, short, stout branches. Skeletal hyphae often dominate in the context and in some cases show a dextrinoid reaction in Melzer's reagent.

Binding hyphae : These in contrast to skeletal hyphae are branched. They may be thick-walled or thin-walled. Sometimes it will be difficult to distinguish scarcely septate, branched generative hyphae and binding hyphae. However, binding hyphae will in most cases taper towards apices while both generative hyphae and skeletal hyphae are of even diameter. The degree and type of branching in binding hyphae differ from genus to genus (Fig. 4).

(j) **Spores :** Spores are important both for generic and specific delimitation although many polypores can be identified without spores. As spore size can vary considerably within a species it is better to measure spores from a spore print whenever possible. The shape and ornamentation of spores differ in different genera and species, although majority of popylores are smooth spored (Fig. 3.).

(k) **Cystidia :** They are sterile structures occurring in the hymenium and/or trama, named Hymenial cystidia and tramal cystidia respectively. Hymenial cystidia originate in the hymenium and are about the same size as the basidia. They are common in monomitic species and have a septum at the base. Cystidia may be thin or thick-walled, smooth or encrusted.

Tramal cystidia: Originate deep in the trama and may be confind to it or may project into the hymenium obliquely. They are thick-walled, often seen in dimitic and Trimitic species.

Cystidioles are smooth,s terile structures having the same size as basidia, but distinguished from the latter by the tapered apex while immature basidia have rounded apex.

Gloeocystidia : are thin-walled with granular or oily content, slightly larger than basidia and projecting above the hymenium. They are rare in polypores (Fig. 5).

(1) **Setae and Setal Hyphae :** These are sterile structures restricted to family Hymenochaetaceae and important in identification of species. All setal organs are thick-walled and dark brown, often more abundant in older parts of the tubes than in the young tissues.

Setae are of two types, **hymenial and tramal;** the former arising in the hymenium and the latter in the trama. Hymenial setae are often conspicuous, projecting beyond basidia; They have a sharp and tapering apex which may be simple and straight, bent to one side or hooked or rarely bifurcate. Setae may be acuminate or ventricose with a swollen part above the base and then tapering to the apex (Fig. 5.)

Tramal setae are mostly embedded in the trama and runs parallel to the tube walls or project obliquely in to the hymenium.

Setal hyphae : The setal hyphea are thickwalled, dark brown, conspicuous structures, which are straight and run parallel to the pore tubes. They are mich bigger than setae, reaching a length of to upto 400 μm, of equal width (up to 25m) tapering only at the apex. The setal hyphae sometimes project into the dissepiments. It is often confused with tramal setae, but true tramal setae are distinctly conical and are widest in the middle part.

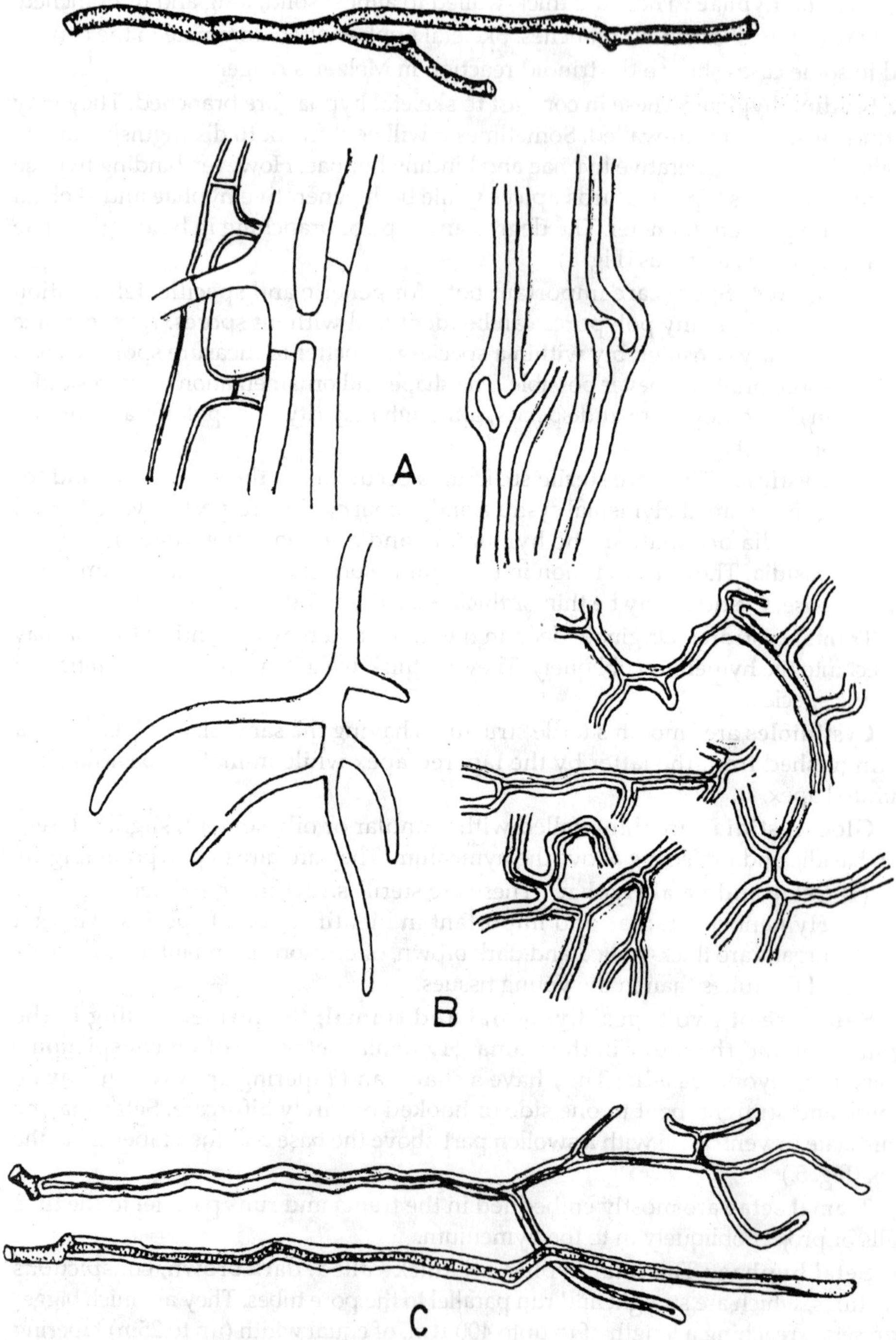

Fig. 4 : Types of *hyphae in polypores.*

A - Generative hyphae thin and thick-walled, simple septate and clamped; B - Binding hyphae, thin and thick-walled; C - Skeletal hyphae, simple and arboriform.

GANODERMATACEAE Donk

SYSTEMATIC ACCOUNT

KEY TO FAMILIES

1. Spores with double wall, an outer hyaline, thin, membraneous exosporium covering an ornamented, thick, brown endosporium, spores round to truncate.................................... **GANODERMATACEAE**

1′. Spores with simple wall, smooth or ornamented, hyaline or brownish..................2

2. Fruitbody in shades of brown, xanthochroic, generative hyphae always simple-septate, hyaline if dimitic, dark brown if monomitic, dark brown setae present or absent.. **HYMENOCHAETACEAE**

2′. Fruitbody white, cream, red, brown or black; generative hyphae simple-septate or with clamps; usually not xanthochroic, if xantho chroic generative hyphae clamped; if mono mitic never with brown generative hyphae; setae never present .... **POLYPORACEAE**

The family Ganodermataceae Donk is characterised by species with spores which are invariably double walled with an inner verrucose to ornamented, thickened and usually coloured wall over which there is a thin hyaline outer wall. It was the unique structure of the spore wall exposed by Atkinson (1908a,b) and Coleman (1927) which induced Donk (1948) to establish this family. Later, Furtado (1962), Heim (1962) and Perreau-Bertrand (1967, 1973) confirmed this spore morphology by means of electron microscopy.

Patouillard (1900) included *Ganoderma* Karst. and *Elfvingia* Karst. in this 'serie des placodes' (tribe 'porohydnes', sub-tribe 'pores', group 'Les Fomes') of the Aphyllophorales. Murrill (1905) erected another genus, *Amauroderma* which he included in the heterogeneous tribe Fomiteae along with *Ganoderma* and *Elfvingia.* Heim (1962) concluded from his studies that the spore wall showed considerable uniformity in the various species he studied and because of this uniformity Patouillard's comprehensive conception of *Ganoderma* (inclusive of *Amauroderma*) was correct and it should be maintained merely as a member of Patouillard's 'serie des placodes'.

The genera included in this family which varies according to different authors are *Amauroderma, Elfvingia, Ganoderma, Haddowia, Humphreya* and *Magoderna.* Steyaert (1972) while erecting the last three genera considered them as a mixture of *Amauroderma* and *Ganoderma.*

According to Donk (1964) a number of polyporaceae such as *Perenniporia* Murr., *Truncospora* Pilat ex Pilat and a few others remind one of *Ganoderma* by their truncate

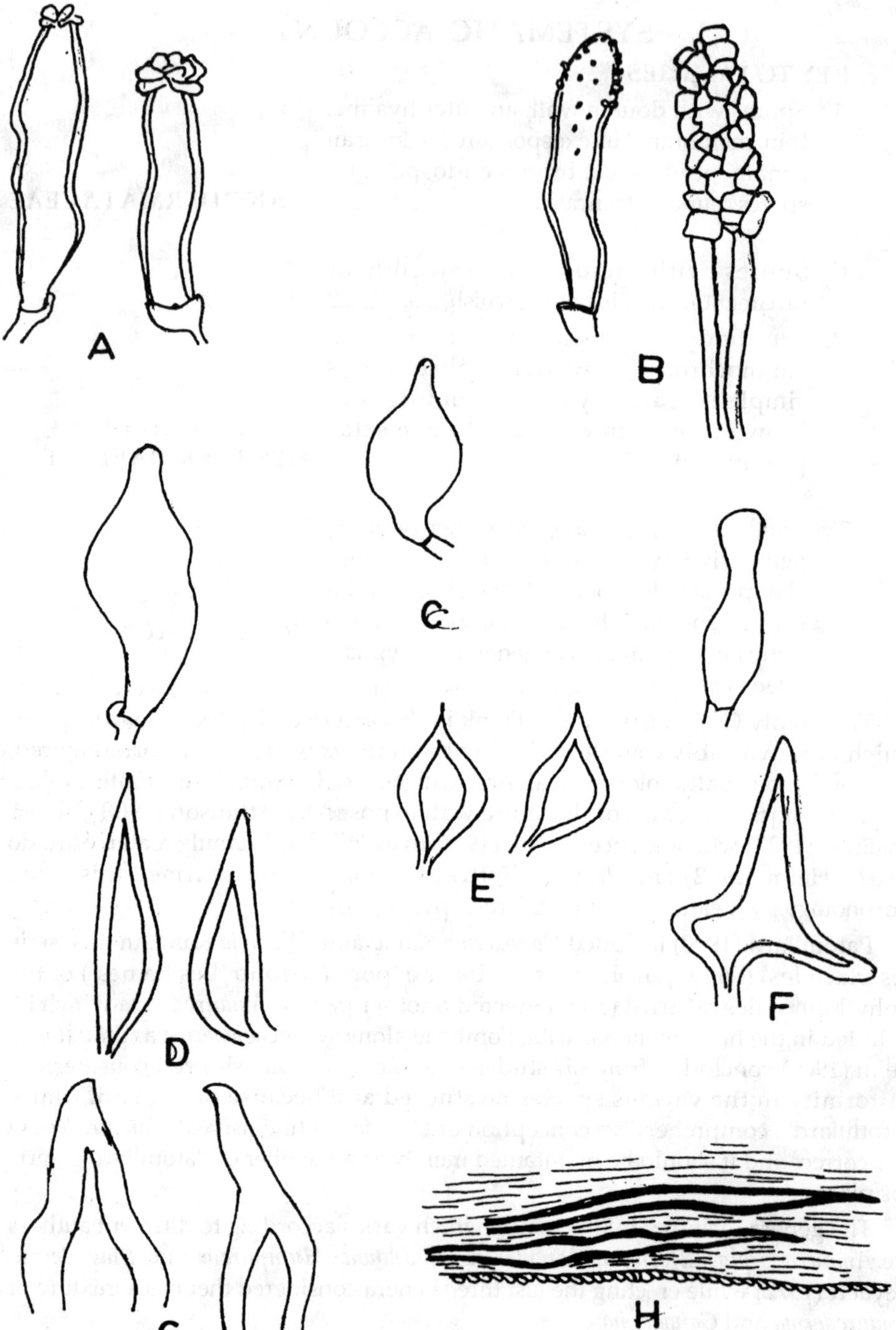

Fig. 5 : Types of *Cysitidia* A-C; *Setae and setal hyphae* D-H;

A & B - Eencrusted cystidia; C- Thin-walled cystidia of different shapes; D - Acuminate setae; E - Ventricose setae; F-Setae with laterel swelling; G -Hooked Setae; H- Setal hyphae in trama.

spores but the spores of these differ in the inner layer which lacks colour as well as ornamentation and is relatively much thinner. Moreover, the truncation of the spores in this group is of an extremely different nature from that of *Ganoderma.* In the pseudoganodermatoid spores the outer wall strongly absorbs cotton blue and strand out as a more prominent layer while in ganodermatoid spores the outer wall is hyaline and noncyanophilic.

Cunningham (1965) included *Ganoderma* under the tribe Ganodermae, *Amauroderma* under the tribe Polyporeae and *Elfvingia* under the tribe Fomiteae, the former two belonging to the sub-family Polyporoideae and the latter to sub-family Fomitoideae in Polyporaceae. He based his classification mainly on the hyphal system and presence of clamps than on spore characters.

Pegler (1973b) accepted only *Amauroderma, Ganoderma* and *Elfvingia* under the family Ganodermataceae and considered genera such as *Dendrophagus* Murr., *Friesia* Lazaro, *Tompophaqus* Murr. and *Whitfordia* Murr. as synonyms to his accepted genera.

Ryvarden & Johansen (1980) considered *Elfvingia* as a synonym of *Ganoderma* while recognising *Amauroderma, Ganoderma, Haddowia* and *Humphreya* as separate genera in the family Ganodermataceae.

HYMENOCHAETACEAE Donk

Hymenochaetaceae Donk is a well marked family of Aphyllophorales characterised by species having setate and/or dark coloured context becoming irreversibly black in KOH solution and clampless generative hyphae. The absence of clamps emphasized by Donk (1948) when establishing the family was supported by Corner (1950) who opined that "the absence of clamps so distinguishes the Xanthochroic series.........that their presence is proof that a species does not belong". The claim of Nobles (1958) that "very rare clamp connections......have been observed in a number of species" has been questioned by Donk (1964). Most species are easily recognised by the presence of setae which are believed to be unique in the Hymenomycetes (Donk, 1964).

The naturalness and homogenity of this group is supported by Nobels' study of the cultural characters of about forty polyporoid species. She reported that with one exception, all of them produced extracellular oxidase in culture and were associated with white rot (Nobles, 1958).

Patouillard (1900), who first recognized this as a separate group, kept them in his 'Serie des Igniaires' which Donk (1933, 1948) accepted in a somewhat amended form by adding Patouillard's 'Serie des Asterostromes'. Thus amended, this family contained members with various types of hymenial configuration. Imazeki & Toki (1954) divided this group into two families viz. Hymenochaetaceae and Mucronoporaceae Imaz. & Toki.

The family Hymenochataceae as envisaged by Donk (1964) consisted of three sub-families viz. Asterostromatoideae Donk., Hymenochaetoideae Donk and Vararioideae Donk with a total of eighteen genera. This included seven genera with non-poroid hymenophore. All genera with poroid hymenophore viz. *Aurificaria* Reid, *Coltricia* S.F. Gray (*Polystictus* Fr. sensu strico), *Coltriciella* Murr., *Cyclomyces* Kunze ex Fr., *Flaviporellus* Murr., *Inonotus* Karst., *Mucronoporus* Ell. & Ev. (*Onnina* Karst.), *Phaeolus* Pat., *Phellinus* Quel., *Phylloporia* Murr. and *Xanthoporia* Murr. belonged to the sub family Hymenochaetoideae which also included the non-poroid genera, *Asterodon* Pat., *Clavariachaete* Corner, *Hydnochaete* Bres, and *Hymenochaete* Lev.

Cunningham (1965) who recognized only thirty five genera altogether in Polyporaceae did not accept a separate family Hymenochaetaceae but five genera considered by others as belonging to this family have been distributed in two tribes Coltriceae (*Coltricia, Cyclomyces, Inonotus*) and Phellineae (*Fuscoporia* Murr., & *Phellinus Quel.*) in the sub family Fomitoideae.

Pegler (1973a) recognised eighteen genera in Hymenochaetaceae, distributed in three sub families viz., Asterostromatoideae, Hymenochaetoideae and Vararioideae. However, all the genera in his (Pegler, 1973a) Hymenochaetaceae are not the same as those of Donk (1964). For example, Pegler (1973a) considered *Xanthoporia* as a synonym of *Inonotus* while *Flaviporellus* and *Phylloporia* recognised by Donk (1964) were not accepted. At the same time, Pegler, (1973a) recognises *Pycnoporellus* Murr., *Cycloporus* (Pat.) Murr. and *Cryptoderma* Imaz. which Donk (1964) had not included in Hymenochaetaceae. In his key to world genera of polypores, Peglar (1973b) has accepted only 12 genera including *Phylloporia* in Hymenochaetaceae while *Vararia* Karst. was put in a separate family Lachnocladiaceae Reid.

POLYPORACEAE Corda

The term polyporaceae has been used with different connotations varying from a heterogeneous assemblage of poroid species to the family Polyporaceae sensu stricto, the latter including only poroid genera not included in other families of Aphyllophorales. According to Donk (1964), whose contribution to classification of Aphyllophorales is outstanding, poroid genera are distributed in about ten families of this order.

Until quite recently, the term Polyporaceae was used in a broad and artificial sense to the large number of poroid species. This is borne out by Corner's (1953) observation that "…….the family is a grade, or artificial assemblage of diverse affinity, at a common level of developing tubes on the underside of the fruitbody. Some are related with clavarioid genera, others with stereoid, hydnoid and, even agaricoid". He also remarked that " indeed, because of this heterogeneous origin and because of their very intricate construction I believe that the natural classificaton of the polypores will be the hardest problem in the systematics of Basidiomycetes."

Up to the time of Patouillard, most families in the higher Basidiomycetes were exclusively based on the nature of the hymenophore, and a species with tubular hymenophore was placed in Polyporaceae. Fries (1874) treated polypores in the order 'Polyporei' under the class 'Hymenio effigurato' and directly classified them into genera. He Put *Lenzities* Fr. in the order 'Agaricini' but *Boletus* Dill. ex Fr. was included in 'Polyporei'. Patouillard (1900), who deviated from the Friesian tradition of classification, kept all the poroid species in the sub tribe 'Pores' of the tribe 'Porohydnes' of his 'Aphyllophoraces'. Patouillard's Aphyllophoraces' roughly corresponds to four out of six orders of Fries viz. Polyporei, Hydnei, Thelephorei and Clavariei.

Donk (1960) defined Polyporaceae as containing "fungi which Fries called or would have called polypores as far as they are homobasidious but with the exclusion of the genera referable to the 'Cyphellaceae', the 'Boletaceae' and the 'Meruliaceae'.

According to Donk (1964) whose system of classification was derived from Patouillard's, the poroid genera comes in more than ten families of which Polyporaceae is only one. However even Donk's family Polyporaceae is "merely a container of all polypores not (yet) assigned to other families" and "is still the biggest family with a large number of genera of doubtful affinity".

Several attempts have been made from quite early times to divide Polyporaceae into smaller families and one result of these efforts was that some members of this family were assigned to the *Agaricales.* Giving much importance to hymenial configuration, polyporaceous genera with more or less typically lamellate hymenophore such as *Lenzities* and some of the tubulate genera such as *Favolus* (P. Beauv) Fr. were shifted to agarics (Fries, 1825, 1874; Patouillard, 1890). Overeem & Weese (1924) opined that certain typical species of *Polyporus* sensu stricto were related to agarics. Overeem (1925) opined that *Polyporus* as emended by him was agaricaceous and close to the species of *Pleurotus* (Fr.) Kumm. Corner (1953) also supported this view based on his observations on the similarity in hyphal system of *Pleurotus* and *Polyporus squamosus* (Huds.) Fr. More recently several authors (Locquin, 1957; Singer, 1962; Kreisel, 1969) have transferred *Polyporus* to *Agaricales,* but Donk (1964) criticized these transfers on the ground that they were not based on redefinition of the generic and family characters. Donk (1964) therefore preferred to maintain the artificial family Polyporaceae in a broader sense among Aphyllophorales unti better solutions were found. Singer has, however, in his later works (Singer, 1975, 1986) while defining family and genera, included *Polyporaceae* as a family in the order *Agaricales.* His *Polyporaceae* (in a restricted sense) consists of *Polyporus* Mich. ex. Fr., *Pseudofavolus* Pat. and *Mycobonia* Pat. under the tribe Polyporeae and *Lentinus* Fr. and related forms under the tribe *Lentiniae.*

Many of the modern workers (Pegler, 1973 a, b; Ryvarden & Johansen, 1980; Gilbertson & Ryvarden, 1986) use the term Polyporaceae in the sense used by Donk (1964), and it is in this sense that the term has been used in the present work.

GANODERMATACEAE Donk

Bull. Bot. Gard. Buitenz. 3, 17 : 474, 1948

Fruitbody annual or perennial, sessile or stipitate. Hymenophore one sided, tubulate, often stratified; pores minute to small, almost regular, edges of dissepiments sterile, context pallid to dark brown or purplish. Hyphal system trimitic; generative hyphae with clamps; skeletal hyphae branched at extremities; binding hyphae usually rare. Cystidia and setae none. Basidia rather short, swollen, 4-spored. Basidiospores globose to ellipsoid or truncate (in *Ganoderma*), with double wall, an outer very thin, hyaline, membrane-like exosporium covering an ornamented thick-walled and (often) brownish endosporium

Type genus : *Ganoderma* Karst.

KEY TO GENERA

1. Fruitbody sessile to stipitate, epixylous or terrestrial, if stipitate never with black zone in stipe and context; spores distinctly truncate.. *Ganoderma*

1′. Fruitbody laterally or centrally stipitate; stipe and pileal context with black zones; spores round.................................. *Amauroderma*

AMAURODERMA Murr.

Bull. Torrey Bot. Club 32 : 366, 1905.

Fruitbody solitary or gregarious, terrestrial, annual to perennial, mesopodial to pleuropodial, coriaceous to woody, orbicular, concave to slightly infundibuliform; blackish or dark brown, finely velutinate to glabrous, wrinkled radially. Stipe usually black, velutinate, slender and long, if hollow then extending into the pileal context getting dissected into smaller tubes and running radially. Pore surface pale-coloured, pores minute, angular to round. Context plae to dark brown. Hyphal system di-to trimitic; generative hyphae either simple-septate or nodose-septate; binding and skeletal hyphae pale-coloured. Cystidia absent. Basidiospores hyaline to pale yellow, sub-globose to cylindrical, large (6-17 µm long), bitunicate with the inner wall finely asperulate and always uniformly thick.

Type species : *Fomes regulicolor* Berk. ex Cke. [=*Amauroderma schomburgkii* Mont. & Berk.]

Only one species has been collected from Kerala.

AMAURODERMA RUGOSUM (Bl. & Nees ex Fr.) Torrend

Broteria (Ci. nat.) 18 : 127, 1920. Fig. 6a - c and i; Pl. fig.

Polyporus rugosus Bl. & Nees, Nova Acta 13 : 21, 1826.

P. rugosus Bl. & Nees ex Fr., Elench. Fung. 1 : 74, 1828.

Fomes rugosus (Bl. & Nees ex Fr.) Cke., Grevillea 13 : 117. 1885.

Ganoderma rugosum (Bl. & Nees ex Fr.) Pat., Bull. Soc. Mycol. Fr. 5 : 68, 1889.

Amauroderma rugosum (Nees) Bose, Ann. Mycol. 35 : 131, 1937, nomen nudum.

Fruitbody annual to perennial, terrestrial, solitary, mesopodial, orbicular, applanate, woody hard, 2-5 cm in diam., 4-8 mm thick. Pileus surface dark brown (6F8) or dark grey (1F1) to black, concentrically and closely zonate, radially sulcate or wrinkled, glabrous; margin deflexed, entire. Stipe sallow (4D3) or golden grey (4C2) at the apex., dark brown (6F8) to black in other areas; often contorted towards the base, otherwise cylindrical, finely velutinate, becoming glabrescent, hollow and extending into the pileal context getting dissected into smaller strands that run radially; inside tissue yellowish brown delimited by a black layer towards the periphery and hollow mid portion; 5-12cm long and to 5mm thick. Pore surface sallow (4D3) to golden grey (4C2) or yellowish olive (4D4), blackish when bruised; smooth and even; pores small and not visible to naked eye, never decurrent, absent near margin, angular to round, regular, 5-7 per mm; pore mouth 110-125 µm wide;

dissepiments 50-70 μm thick; pore tubes concolorous with the pore surface or brownish yellow, up to 3.5mm long, sunk into equal depths, straight. Context pale yellow (2A3) while fresh, ivory (4B3) when dry, homogeneous, up to 3mm thick, of intertwined hyphae, compactly arranged.

Hyphal system di-to trimitic. Generative hyphae hyaline, thin-walled, branched, simple-septate, clamps not observed, up to 2.5 μm in diam. Skeletal hyphae pale-coloured, thick-walled with narrow lumen or solid, well branched at extremities, branches contorted, non-septate, central stem cylindrical and straight, 3-7.5 μm in diam. Basidiospores (8.5) 9.5-12.5 × 7.5-9.5(10) μm, exosporium smooth, endosporium with fine projections (asperulate). Pileal cuticle consisting of gelatinized, thick-walled, parallel skeletal hyphae arising obliquely, up to 7.5 μm in diam.

SPECIMEN EXAMINED : On the ground near bamboo bushes, from a forest patch, Vazhachal, Sholayar forests, Trichur Dist., Herb. CALI No. G27, 29-8-1983; on ground, from an inhabited area, Pazhayangadi, Cannanore Dist., Herb. CALI No. G381, 7-11-1987; on ground, Botanical garden, Calicut University Campus, Malappuram Dist., Herb. CALI No. F988, 28-9-98.

GEOGRAPHICAL DISTRIBUTION : India, Sri Lanka, Philippines, Perak, Australia Africa, Brazil, Guiana.

REMARKS : *Amauroderma rugosum* was described earlier from India by Bose (1937a). Banerjee (1947) and Bakshi (1971), based on collections from north-east regions. Bakshi reports *A. rugosum* with stipe up to 22 cm long. In New Zealandean sporophores Cunningham (1965) reports parallel hyphae in the context. The description of this species available at Boger and Leiden Herbaria by Steyaert (1972) and of the East African sporophores by Ryvarden & Johansen (1980) agree closely with the present collection. It is noteworthy that some of the collections of a *A. rugosum* was from areas close to bamboo bushes as was the earlier Indian collection from West Bengal (Bose, 1937a).

GANODERMA Karst.

Rev. Mycol. 3 : 17, 1881.

Fruitbody terrestrial or lignicolous annal or perennial, stipitate or sessile. Pileus surface with a thick dull or shiny and laccate crust, a thin hymenoderm of clavate end cells. Pore surface cream-coloured, brown on bruising, pores regular, 4-7 per mm, tube layers single or stratified, pale to purplish brown. Stipe when present central or lateral. Context cream-coloured to dark purplish brown, soft and spongy to firm-fibrous. Hyphal system trimitic; generative hyphae with clamps; skeletal hyphae hyaline to brown, non-septate, often with long, tapering branches. Basidia broadly ellipsoid, tapering abruptly at the base. Cystidia absent. Basidiospores broadly to

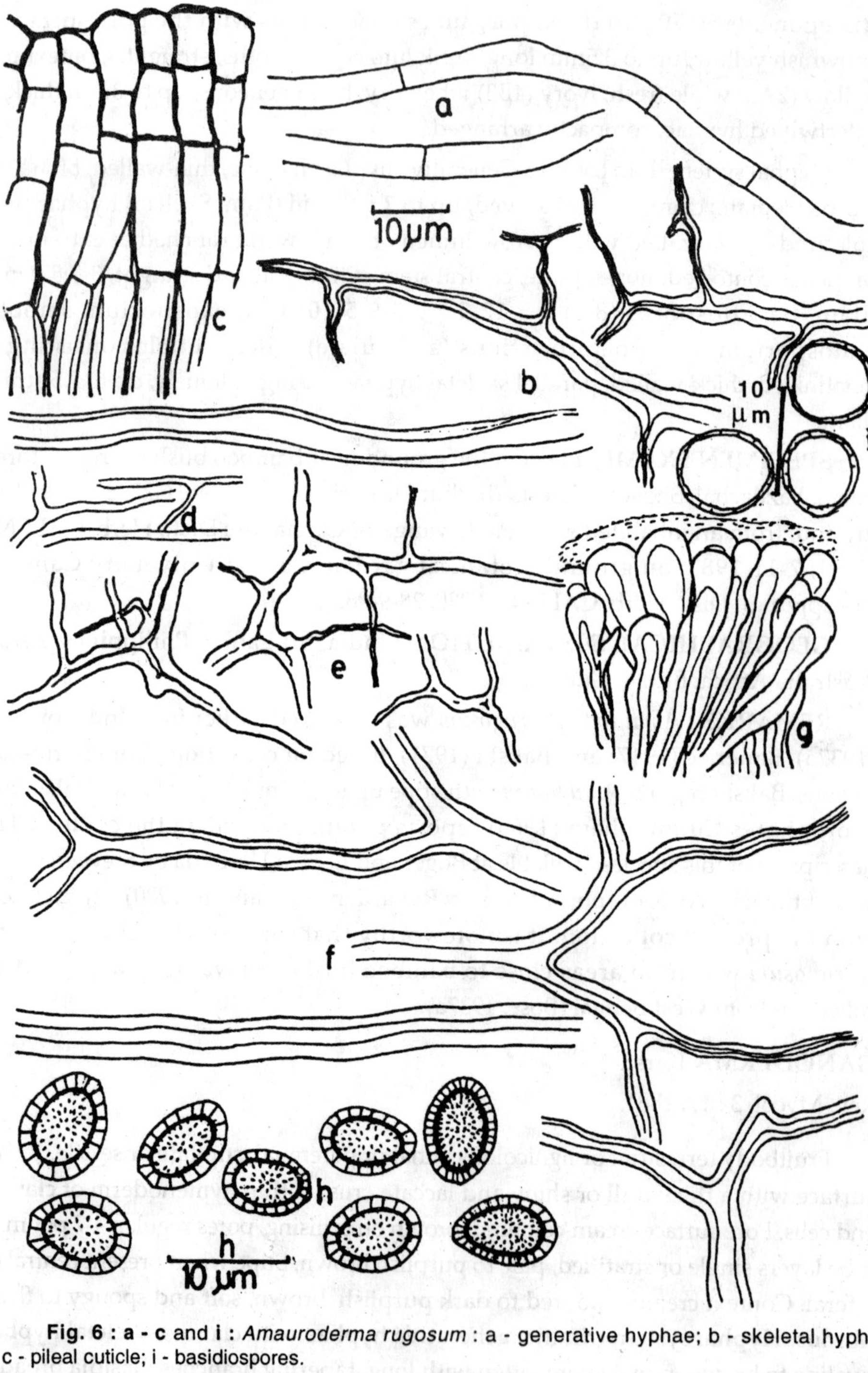

Fig. 6 : a - c and i : *Amauroderma rugosum* : a - generative hyphae; b - skeletal hyphae; c - pileal cuticle; i - basidiospores.

d - h *Ganoderma australe* : d - generative hyphae; e - binding hyphae; f - skeletal hyphae; g - pileal cuticle; h - basidiospores.

narrowly ellipsoid tapering abruptly with a truncate apex and apical germ pore, wall two-layered, the endosporium brown and separated from the hyaline exosporium by interwall pillars, negative in Melzer's, 7-20 µm long. Causing uniform or mottled white rot of dead and living hardwoods and conifers.

Large, cosmopolitan genus but most species in the tropics.

Type species : *Ganoderma lucidum* (Curt. ex Fr.) Karst.

KEY TO SPECIES

1. Fruitbody stipitate; pileus surface, laccate, reddish brown or yellowish; cutis a hymenoderm of clavate hyphal endings, 12-15 µm broad .... *G. lucidum*

1'. Fruitbody sessile; pileus surface not laccate and shiny, brownish or darker; cutis of narrowly clavate elements, elements, 5-10 µm broad .... *G. australe*

GANODERMA AUSTRALE (Fr.) Pat.

Bull. trimistr. Soc. Mycol. fr. 5 : 71, 1889.

Fig. 6. d - h

Polyporus australis Fr., Elench Fung. p. 108, 1828.

P. tornatus Pers., Freyc. Voy. Uranie p. 173, 1827.

Ganoderma tornatum (Pers.) Bres., Hedwigia 53 : 55, 1912.

Fomes oroflavas Lloyd, Mycol. Letter 48 : 8. 1913.

Elfvingia browii Murr., West Polyp. p. 29, 1915.

Ganoderma oroflavum (Lloyd) Humph., Philipp. J. Sci. 455 : 503, 1931.

G. applanatum (Pers.) Pat. var. laevisporum Humph. & Leus, Ibid. 45 : 533, 1931.

G. applanatum (Pers.) Pat. var. philippinse Humph. & Leus, Ibid. 45 : 535, 1931.

G. applanatum (Pers.) Pat. var. tornatum (Pers.) Humph. & Leus forma macrosporum Humph. & Leus, Ibid. 45 : 551, 1931.

G. brownii (Murr.) Gilbn., Mycologia 53 : 505, 1961.

Fruitbody annual to perennial, pileate, sessile or very rarely sub-stipitate, attached with a broad lateral base, semi-circular to dimidiate, rarely irregular, applanate to conchate, hard corky to woody while fresh, tough but light in weight when dry; 4.5-26 × 4-16 × 1-4.5 cm. Pileus surface Sahara (6C5) to brownish orange (7C5) or brownish grey (6F2) to black with deposits of sunburn (6D5) or cocoa brown (6E6) powder in grooves, sometimes linoleum brown (5E7) to darker, smooth or rough, uneven, tuberculate, concentrically sulcate, radially ridged, semi-glabrous, heavily crusty; margin white, to 1 cm broad, entire and rounded. Pore surface reddish white (10A2) to almost ash grey (1B2) or smoke grey (3C2), sometimes eye brown (7F6) to greyish brown (7E3) or dark blonde (5D4) while fresh, yellowish white (2B2) to yellowish grey (2C2) when mature or dry, sometimes reddish blonde (5C4) towards margin, even and smooth; pores not extending to the margin, small and visible to naked eye, round, almost regular, 4-6 per mm; pore mouth (85) 100-125(175) µm wide; dissepiments (50) 65-100(120) µm thick; pore tubes raw umber (5F8) to almost concolorous with the pore surface, older layers often eye brown (7E6), 5-8mm in each stratum, seldom 12-17 mm long. Context dark brown (7E7) to rust brown (6E8) or cognac (6E7) with white mycelial strands (up to 1mm in diam.) transversely running in between, rarely homogeneous, fibrous, shiny, (0.2) 0.5-0.7 mm thick, of compact, intertwined hyphae.

Hyphal system trimitic. Generative hyphae hyaline, thin-walled, sparsely branched, septa or clamps not observed, 2-3 µm in diam. Binding hyphae hyaline to brownish, thickwalled, branchings of the Bovista-type, twisted and flexuous, ends attenuated, common in the white mycelium of the context, (2) 3-5 µm in diam. Skeletal hyphae hyaline soon becoming brown, thick-walled with obterate lumen or solid, extremities with few branchings, 5-7 µm in diam. Pileal cuticle a trichodermal palisade of thick-walled inflated cells, (5)7.5-10 µm broad; hyphal branchings held together by melanoid deposits, spores and other crystalline bodies to form a crusty layer. Basidiospores golden brown to brownish yellow, thick-walled, broadly ellipsoid to truncate, exosporium hyaline, thin and smooth, endosporium coarsely ornamented, (6)6.5-8-9.5(10) x 4.5-5-5.5)(6) µm; spore print sunburn (6D5).

DECAY : White rot.

SPECIMEN EXAMINED : Decaying logs of *Buchanania axillaris* (Desr.) Ram. (ANACARDIACEAE), (Vengoli hills, Tunacadavu, Parambiculam forests, Palghat Dist., Herb. CALI No. G201, 11.7. 1985; on dead trunk of *Cocos nucifera* Linn. (PALMAE) Botanical Garden, Calicut University Campus, Malappuram Dist., Herb. CALI No. G322, 25-6-1986; on the trunk of a living *Ficus* sp. (MORACEAE), Mundakai, Wyanaad Dist., Herb. CALI No. G123, 22-9-1984; on dead trunk of *Syzygium cuminii* (Linn.) Skeels (MYRTACEAE), Chattapuram, Palghat Dist., Herb. CALI No. G285, 8-4-1983; on decaying *Toona ciliata* Roemer (MELIACEAE), Kanoth Reserve Forets, Wyanaad Dist., Herb. FRI No. 4242, K. Bagchee, 25-1-1946; on dead logs of *Mesua nagassarium* (Burm. f.) Kosterm. (CLUSIACEAE), Perumpara (W), Malakkapara, Sholayar forests, Trichur Dist., Herb. CALI No. G174, 10-6-1985; on unidentified forest logs, Muthanga Reserve Forests, Sultan Battery, Wyanaad Dist., Herb. CALI No. G195, 20-6-1985; Perumpara (W), Malakkapara, Sholayar forests, Trichur Dist., Herb. CALI No. G349, 30-8-1986; Velangad, Calicut Dist., Herb. CALI No. G271, 3-1-1986.

GEOGRAPHICAL DISTRIBUTION : A tropical and sub-tropical species.

REMARKS : This is the first detailed description of this species from India although Bakshi (1971) had mentioned *G. australe* under the imperfectly known polypores. Steyaert who received a collection of *Ganoderma* from Kerala through Bakshi (FRI 4242) described it as *G. tornatum* (Pers.) Bres (Steyaert, 1972). Though the present collection agrees well with the description of *G. tornatum* by Steyaert (1967, 1972), Ryvarden & Johansen (1980) allocates nomenclatural priority to *G. australe* (Fr.) Pat. The present collection agrees well with the description of *G. australe* by Bonderstev (1953), Bourdot & Galzin (1927), Ireneia (1986), Murrill (1915) and Ryvarden & Johansen (1980). The New Zealandean sporophores (Cunningham, 1965) have wider pore mouths and dissepiments as well as bigger spores. Rea (1922) and Steyaert (1975) also reports bigger spores.

GANODERMA LUCIDUM (Curt. ex Fr.) Karst.

Rev. Mycol. 3, 9 : 17, 1881.

Fig. 7a - f; Pl. Fig

Boletus lucidus Curt., Fl. Iondin 2, pl. 224, 1781 *Polyporus lucidus* (Curt.) ex Fr., Syst. Mycol. 1 : 353, 1821.

Boletus laccatus Timm., Fl. megalop. Prodr. p. 269, 1788.

Polyporus Iaccatus (Timm.) Pers., Mycol. Eur. 2 : 54, 1825, non Kalch., 1885.

P. japonicus Fr., Epicrisis p. 442, 1838.

Fomes japonicus (Fr.) Sacc., Syll. Fung. 6 : 156, 1888.

Placodes lucidus Quel., Fl. Nycol. p. 399, 1888.

Ganoderma pseudoboletus (Jacq.) Murr., Bull. Torrey Bot. Club 29 : 602, 1902.

Ganoderma sessile Murr., Bull. Torrey Bot. Club 19 : 604, 1902.

Polyporus polychromes Copeland, Ann. Mycol. 2 : 507, 1904.

Ganoderma subperatum Atk., Gaz. 46 : 337, 1908.

Polyporus incrustans Lloyd, Synop. Stip. Polyp. p. 102, 1912.

P. metalicus Lloyd, Mycol. Notes 65. : 1099, 1921. 1932, non Bourd. & Galz., 1928.

Ganoderma laccatum Pat. in Bres., Ic. Myc. 21 : 1004.

Fruitbody epixylous or attached to underground roots, annual, centrally or laterally stipitate, rarely substipitate, solitary or imbricate, orbicular to semi-circular, sometimes variously lobate to appear like a flower, corky in consistency, woody when dry; 5-15 × 4-8 × 0.75-1.5 cm. Pileus surface oxblood red (9E7), reddish brown (9D7), violet brown (11E4), caput mortuum (8F7) to almost black, oak brown (5D6) to terra cota (7D7) or pastel yellow (2A4) towards margin, colour slightly varies in different sporophores; glabrous, smooth, shiny, laccate, faintly crusty and concentrically plicate; margin thick, smooth, entire, rounded. Stipe fox brown (8D7) or black, glabrous, cylindrical to flattened, reduced or up to 8 cm long, 0.8-2.5 cm thick; stipe-tissue leather brown (6E6) at the centre to Pompeian yellow (5C6) towards periphery, fibrous solid. Pore surface pale grey (IBI) or cream (4A3), sometimes golden blonde (5C4) to greyish orange (5B3), becoming brownish on bruising, even and smooth; pores almost visible to naked eye, extending to the margin, slightly decurrent, round to somewhat angular, 4-5(6) per mm; pore mouth 100-150 μm wide, dissepiments 50-75 μm thick; pore tubes concolorous with the pore surface, 0.3-0.8cm long, straight, sunk into equal depths., Context cognac (6E7), maize (4A6) coloured towards margin, shiny, homogeneous, of intertwined hyphae compactly arranged.

Hyphal system trimitc. Generative hyphae hyaline, thin-walled, branched, septate with clamps, 2.5-4 μm in diam., rare in context. Binding hyphae pale-coloured to slightly yellowish brown, thick-walled with narrow but distinct lumen, branchings of the bovista type, extremities tortuous and attenuated, usually well branched at the central portion, 2.5-4(5) μm in diam. Skeletal hyphae yellowish brown, thick-walled with broad lumen, unbranched or with few terminal branches, 3-6 μm in diam. Pileal cuticle a typical hymeniderm consisting of end cells of hyphae anticlinal to the pileal surface, clavate, thick-walled, about 45-50 × 12.5-15 μm, compactly arranged, with encrustation. Basidiospores yellowish brown, thick-walled, truncate, outer wall hyaline, inner wall thick, finely verrucose and colored, nonamyloid, 8-10 × 5-6μm. Proterospores pale to cream-coloured, round, thick-walled, verrucose, sometimes slightly split, nonamyloid, 14-17 μm in diam.

DECAY : White spongy rot. Initially decayed wood with white mycelial flakes, later collapsing into white mass.

SPECIMEN EXAMINED : On decaying logs of *Anacardium occidentale* Linn. (ANNACARDIACEAE), Botanical Garden, Calicut University Campus, Malappuram Dist., Herb. CALI No. G75, 23-7-1984; on the exposed roots of *Delonix regia* (Hook.) Rafin (CAESALPINIACEAE), Kannara, near Peechi, Trichur Dist., Herb. CALI No. G18, 12-7-1983; on the ground, around the base of *Acacia auriculiformis* A. Cunn ex Benth. (MIMOSACEAE), Calicut University campus, Malappuram Dist., Herb. CALI No. G383, 20-8-1987; on the roots of *Cocos nucifera* Linn. (PALMAE), Botanical Garden, Calicut University campus, Malappuram Dist., Herb. CALI No. G94, 16-9-1983; on the base of the trunk of living *Hydnocarpus pentagyna* (Buch. - Ham.) Oken (FLACOURTIACEAE), Nedumgayam, Nilambur forests, Malappuram Dist., Herb. CALI No. G21, 19-7-1983; on the base of the trunk of living *Theobroma cacao* Linn. (STERCULIACEAE), Alpara, Kannara, Peechi, Trichur Dist., Herb. CALI No. G15, 12-7-1983; on unidentified logs occurring in a semi-deciduous forest, Mundakai, near Meppadi, Wyanaad Dist., Herb. CALI No. G120, 22-9-1984; on decaying logs. Perumpara (East), Sholayar forests, Malakkapara, Trichur Dist., Herb. CALI No. G330, 29-8-1986; on unidentified forest logs, Kanoth Range, Wyanaad Division, Wyanaad Dist., Herb. FRI No. 4255, K. Bagchee, 8-4-1946.

GEOGRAPHICAL DISTRIBUTION : Cosmopolitan.

REMARKS : *G. lucidum* has been frequently collected from India and many researchers have studied its sporophore morphology, cultural characters and pathology (Bagchee et. al., 1954 ; Bakshi, 1971 ; Banerjee & Sarkar, 1956 ; Sarkar,

1969 ; Thind et al., 1957 ; Vijayan et. al., 1973). Bakshi (1971) and Thind et al. (1957) report *G. lucidum* with wider pores than those of the present collection. Proterospores ('abnormal spores') in *G. lucidum* observed by Bose (1933) was noticed in one of the present collections (G21). The basidiospores of *G. lucidum* described by Bazzalo & Wright (1982), Bonderstev (1953), Bourdot & Galzin (1927), Ireneia (1986), Lowe (1934), Lowe & Gilbertson (1961), Rea (1922), Ryvarden (1976 a), Ryvarden & Johansen (1980) and Steyaert (1972) are slightly bigger than those of the present collection. Petersen (1987) reports smaller spores for European species of *G. lucidum.*

HYMENOCHAETACEAE Donk

Bull. Bot. Gard. Buitenz. 3, 17 : 474, 1948

Fruitbody stipitate, pileate to resupinate, annual to perennial, xanthochroic. Pileus surface golden brown to deep sienna brown, occasionally black, tomentose to glabrous, often zoned, smooth or sulcate, occasionally deeply cracked, with or without a distinct crust. Pore surface smooth, hydnoid, poroid or even lamellate. Hyphal system mono-or dimitic; generative hyphea always simple-septate, in monomitic species hyphae mostly hyaline or pale golden brown; skeletal hyphae, if present, golden to rusty brown. Setal hyphae, tramal or hymenial setae absent or present. Basidiospores hyaline to rusty brown, thin-walled, usally negative in Melzer's, dextrinoid in a few species. Mostly saprophytic, more rarely parasitic on living plants, always associated with a white rot.

Cosmopolitan family, richly represented in the tropics.

Type genus : *Hymenochaete* Lev.

4 genera and 22 species have been described here.

KEY TO GENERA

1. Fruitbody stipitate .... *Coltricia*

1'. Fruitbody resupinate to pileate 2

2.' Hyphal system dimitic, skeletal hyphae present, fruitbodies perennial and hard .... *Phellinus*

2.' Hyphal system monomitic, generative hyphae hyaline to rusty brown, fruitbody thin and flexible, a distinct black zone present between an upper tomentum and denser context 3

3. Fruitbodies applanate, thin setae always present, on dead wood.......... *Cyclomyces.*

3.' Fruitbodies convex, setae absent, on living trees .... *Phylloporia*

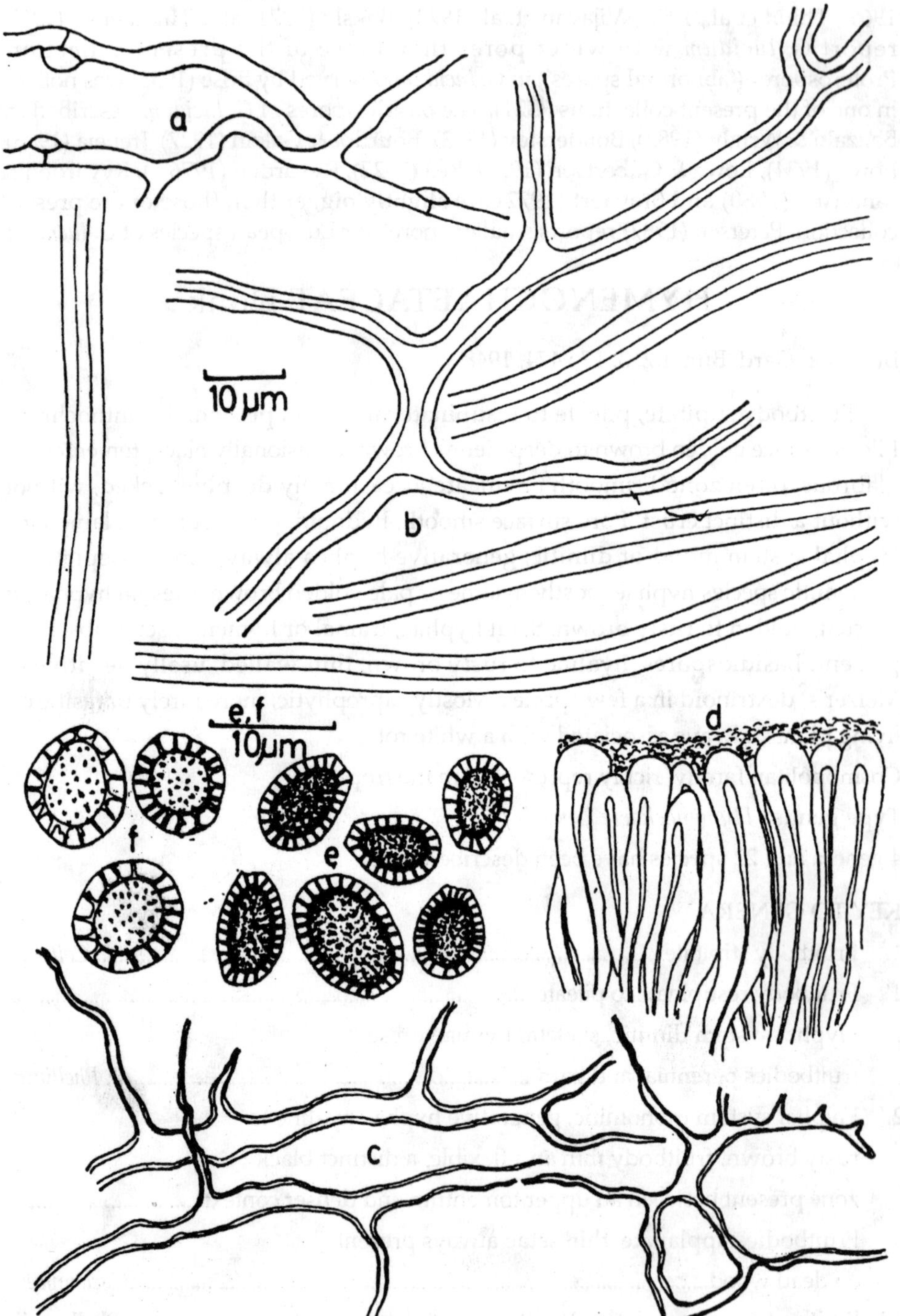

Fig. 7 : a- f : *Ganoderma lucidum* : a - generative hyphae ; b - skeletal hyphae ; c- binding hyphae; d - pileal cuticle; e - basidiospores; f - proterospores.

COLTRICIA S.F. Gray

Nat. Arrang. Brit. Plants 1 : 644 1821.

Fruitbody annual, stipitate with central to lateral stipe, pileus velutinate to hirsute, yellowish to deep rusty-brown or grey with age, mostly concentrically zonate, stipe yellowish to rusty-brown, velutinate to tomentose, pore surface brown, pores angular to round, entire, context yellowish-brown to rusty-brown, often duplex, hyphal system monomitic and generative hyphae hyaline to golden-brown or rusty-brown and with simple septa, setae and setal hyphae absent or present, spores hyaline to light yellowish with slightly thickened walls, smooth, non-amyloid or rarely dextrinoid, mostly on the ground, but also on burnt wood, on dead culms of grasses or buried wood. Cosmopolitan genus.

Only one species collected from this region.

Type species : *Coltricia perennis* (Fr.) Murr.

COLTRICIA FOCICOLA (Berk. & Curt.) Murr.

NORTH AMERICAN FLORA 9 : 92, 1908

Polyporus Focicola Berk. & Curt., J.Linn. Soc. Bot. 10 : 305, 1868

Fig 32. c-e

Fruitbody annual, centrally stipitate, pileus orbicular, shallow infundibuliform,. 5mm to 1.2cm in diam., surface dark brown to cognac (6F8 to 6E8), smooth and satiny, shiny, non-zonate. Stipe central, cylindric, solid, concolorous to pileus surface, 0. 8- 1. 6 cm x 1mm or less, slightly expanded to apex, base slightly bulbous. Pore surface almost concolorous or slightly lighter shade of pileus surface, not decurrent on stipe, margin fertile ; pores entire, circular to polygonal, 1. 5 to 2 per mm, pore mouth wavy ; pore tubes concolorous to pore surface, shallow, less than 1mm deep. Context thin, homogenous, 250-500 μm thick, light brown.

Hyphal system monometic of light brown hyaline hyphae. mostly simple septate clamps rare, branched and thick-walled, 3-3. 7.5 μm broad, Basidia clavate, 20 x 7. 5 μm, basidiospores ellipsoid to broadly ellipsoid 6-8.5 (9.0) x 4.5-5 (5.5) μm, hyaline, thin to slightly thick-walled, smooth, non-amyloid and non-dextrinoid. Setae and setal hyphae absent.

SPECIMEN EXAMINED : On the ground in open grassland, Calicut University Campus, Malappuram Dist. Kerala , India, 21-6-1996. Herb. Cali. No F838.

REMARKS : The present collection comes close to C. *cinnamonea* (Pers.) Murr. in general habit and colour of different parts of the fruitbody but the basidiomes are much smaller in size. The habit and habitat of the two also show differences. The present collection grows at the bottom of shallow soil pits in open grassland, isolated or rarely two together but not in clusters.

CYCLOMYCES Kunze ex Fr.

Linnaea 5 : 512, 1830.

Fruitbody annual, pileate, solitary or imbricate, brownish shaded, semi-circular to flabelliform or sometimes dimidiate with a contracted base, finely pubescent. Pores variable or concentrically lamellate. Context distinctly duplexed, lower part dense, separated from the pileal tomentum by thin black zone. Hyphal system monomitic ; generative hyphae hyaline to brownish and with simple-septa. Setae present, dark brown, acute and thick-walled. Basidiospores hyaline to very pale yollowish, cylindrical to ellipsoid. On deciduous wood. Pantropical genus with four species.

Type species : *Cyclomyces fuscus* Fr.

KEY TO SPECIES

1. Pore mouth rough, pores visible to naked eye, 1 - 2 per mm ; dissepiments dentate and projecting ; setae 5-7 µm broad .... *C. setiporus*

1.' Pore mouth smooth and even, pores not visible to naked eye, 5-6 per mm; dissepiments never projecting; setae 7. 5-10µm broad .... *C. tabacinus*

CYCLOMYCES SETIPORUS (Berk.) Pat.

Essai Tax. P. 98, 1900.

Fig. 8. A-c; Pl., Fig.

Polypours setiporus Berk. , Lond. J. Bot. 6 : 505, 1847.

P. intybaceus Berk., Ibid. 1: 149, 1842, nomen nudum, non Fr., 1838.

P. cichoraceus Berk. ex Fr., Nove Acta 3, 1 : 92, 1851.

Polystictus cichoraceus (Berk.) Fr., Ibid., 3, 1 : 92, 1851.

Hexagonia tabacina Lev. In Zoll., Syst. Ind. Arch. P.15, 1854.

H. arata Berk ., J. Linn. Soc. 16 : 43, 1954.

Favolus transiens Ces., Atti Accad. Sci. Fis. Mat. Napoli 8 : 8,1879.

Polystictus setiporus (Berk.) Cke., Grevillea 14 : 86, 1886.

Cyclomyces cichoraceus (Berk.) Pat., Essai Tax. P. 98, 1900.

Cycloporellus cichoraceus (Berk.) Murr., Bull. Torrey Bot. Club 34 : 468, 1907.

Inonotus setiporus (Berk.) G.H. Cunn., Proc. Linn. Soc. N.S.W. 75 : 242, 1950.

Fruitbody epixylous, annual, sessile, imbricate and confluent, attached with a narrow lateral or central base, centrally attached sporophores drooping like an inverted cup, otherwise semi-circular to flabelliform, lobate towards the margin, reflexed, consistency coriaceous, flexible when fresh as well as when dry, light in weight; 3-6 x 3-6 x 0.3-0.5 cm. Pileus surface yellowish brown (5E8) to rust brown (6E8), sometimes with zonations of allied shades; concentrically striate, striations closer towards margin, radially wrinkled, tomentose, slightly shining; margin thin and entire. Pore surface golden brown (5D7) to yellowish brown (5E8), uneven and rough; pores large and conspicuous, extending to the margin, five to six sided, rarely roundish, 1-2 per mm; dissepiments usually projecting as if irpicoid, 60-120 µm thick; pore tubes concolorous with the pore surface, 2-4 mm long. Context light brown (6D8), dark brown (6F8) towards pileus surface, up to 2mm thick, considerably thinner towards the margin.

Hyphal system monomitic. Generative hyphae yellowish to yellowish brown, slightly thick-walled, branched towards extremities, ends sub-hyaline, septate, 2.5-5 µm in diam; abhymenial hyphae 3.75-6 µm in diam. Setae numerous; yellowish brown, dark brown in KOH, long, acuminate, slightly thick and often bent near the base, 44-55 x 5-7 µm. Basidiospores not observed.

DECAY : White rot, in distinct pockets.

SPECIMEN EXAMINED : On unidentified decaying forest logs, Kaduvakkunnu, Nadugani forests, Nilambur, Malappuram Dist., Herb. CALI No. G54, 28-2-1984; G127, 13-11-1984; G369, 19-4-1987.

GEOGRAPHICAL DISTRIBUTION : Australia, New Zealand, Fiji, Sri Lanka, India, China, Philippines, Malaya, New Caledonia, South America, New Guinea, Semoa Islands, East and West Africa.

REMARKS : *Cyclomyces setiporus* was earlier described as *Polyporus cichoraceus* by Bose (1971) from West Bengal and Coimbatore (Tamil Nadu State). In the setal and macrocharacters the present collection agrees with the earlier descriptions. But Bose (1946) described it as having a trimitic hyphal system with clamps and Bakshi (1971) reported a dimitic hyphal system with clamps. Descriptions of *C. setiporus* from East Africa (Ryvarden & Johansen, 1980) and New Zealand (Cunningham, 1965) agree well with the present collection.

CYCLOMYCES TABACINUS (Mont.) Pat.

Essai Tax. P. 98, 1900. Fig. 8. d-f; Pl., Fig.

Polyporus tabacinus Mont., Ann. Sci. Nat. 3, 3 : 349, 1835.

P. spadiceus Jungh., Verh. Natav. Genootsch. 17 : 54, 1838.

P. microcyclus Lev., Ann. Sci. Nat. 3, 2 : 188, 1844.

Polystictus tabacimus (Mont.) Fr., Nov. Symb. P. 93, 1851.

Inonotus corrosus Murr., Bull. Torrey Bot, Club 31 : 598, 1904.

I.elmerianus Murr., Ibid. 34 : 471, 1907.

Cycloporellus barbatus Murr., Ibid. 35 : 397, 1908.

Polystictus glabrotabacinus Lloyd, Ibid. 74 : 1334, 1925.

P. gilvorigidus Lloyd, Ibid. 74 : 1334, 1925.

Inonotus tabacinus (Mont.) Karst., Rev. Mycol. 3: 19, 1881.

Fruitbody annual or biennial, pileate, sessile, laterally attached with a converging base; densely imbricate, confluent, dimidiate to flabelliform, lobate towards margin, lobes partly overlapping, applanate, thin, coriaceous and flexible while fresh, tough when dry; 4-10 × 3-4.5 × 0.1-0.2 cm. Pileus surface oak brown (5D6) to cognac (6E7); somewhat concentrically striate, faintly pubescent, hairs more concentrated towards the base, almost glabrous towards margin in old specimens;

margin reflexed, rigid and entire. Pore surface bronze brown (5E5) to mustard brown (5E6), smooth and even; pores small, extending almost to the margin, round to somewhat angular, regular, 5-7 per mm; pore mouth 100-125 µm wide; dissepiments 25-50 µm thick; pore tubes concolorous with the pore surface, straight, sunk into equal depths, about 0.5mm long. Context concolorous with the pore surface, dark brown in KOH, with two or three thin black zones running radially, about 1 mm thick.

Hyphal system monomitic. Generative hyphae brownish yellow, thick-walled with distinct lumen, unbranched or rarely branched, simple-septate, width uniform, 3. 75-5 µm in diam. Setae present; yellowish brown, darker in KOH, acuminate, more or less evently tapering, sometimes base curved, 25-35 × 75-10 µm. Basidia and basidiospores not observed.

DECAY : White pocket rot.

SPECIMEN EXAMINED : On unidentified logs, Kerala-Karnataka State boundary, Kasaragod Dist., Herb CALI No. G165, 17-3-1985; Kaduvakkunnu, Nadugai forests, Nilambur, Malappuram Dist., Herb. CALI No. G368, 19-4-1987; Kathiyur Reserve Forest, Wyanaad Dist., Herb. FRI. No. 4256, K. Bagchee, 1-6-1946.

GEOGRAPHICAL DISTRIBUTION : Australia, New Zealand, Fiji, India, Philippines, West Indies, French Antilles, French Guiana, Chile, South America, East Africa.

REMARKS : Bakshi (1971) who described it as *Polyporus tabacinus* attributed wide distribution to this fungus on dead hardwood throughout India. A comparison with Bakshi's (1971) description shows that the pores in his collection are smaller than those of the present collections. Cunningham (1965) and Ryvarden & Johansen (1980) also reports smaller pores in addition to narrower setae compared to author's collections.

PHELLINUS Quel.

Elench. Fung. P. 172, 1886.

Fruitbody annual or perennial, resupinate or pileate, when pileate attached with a broad base. Pileus surface dark brown to black or yellowish grey to ochraceous, hirsute to glabrous, often rimose, sulcate or concentrically striate, radially cracking in older fruitbodies, pores variable, daedaleoid to irregular but usually round and regular. Tubes usually stratified. Context thin or thick, hyphae intertwined or parallel, compactly arranged. Hyphal system dimitic; generative hyphae usually hyaline, thin-walled and narrow, more rarely wider and pale golden brown; skeletal hyphae yellowish or brownish, darker in KOH, prominent. Setal hyphae, tramal setae and hymenial setae present or absent. Basidiospores of variable shapes, hyaline to rusty brown, thin-walled to thick-walled; mostly negative in Melzer's reagent, dextrinoid in a few species. Cosmopolitan genus causing white rot.

Type species : *Polyporus rubriporus* Quel. [=*Polyporus torulosus* Pers.]

18 species have been collected from Kerala.

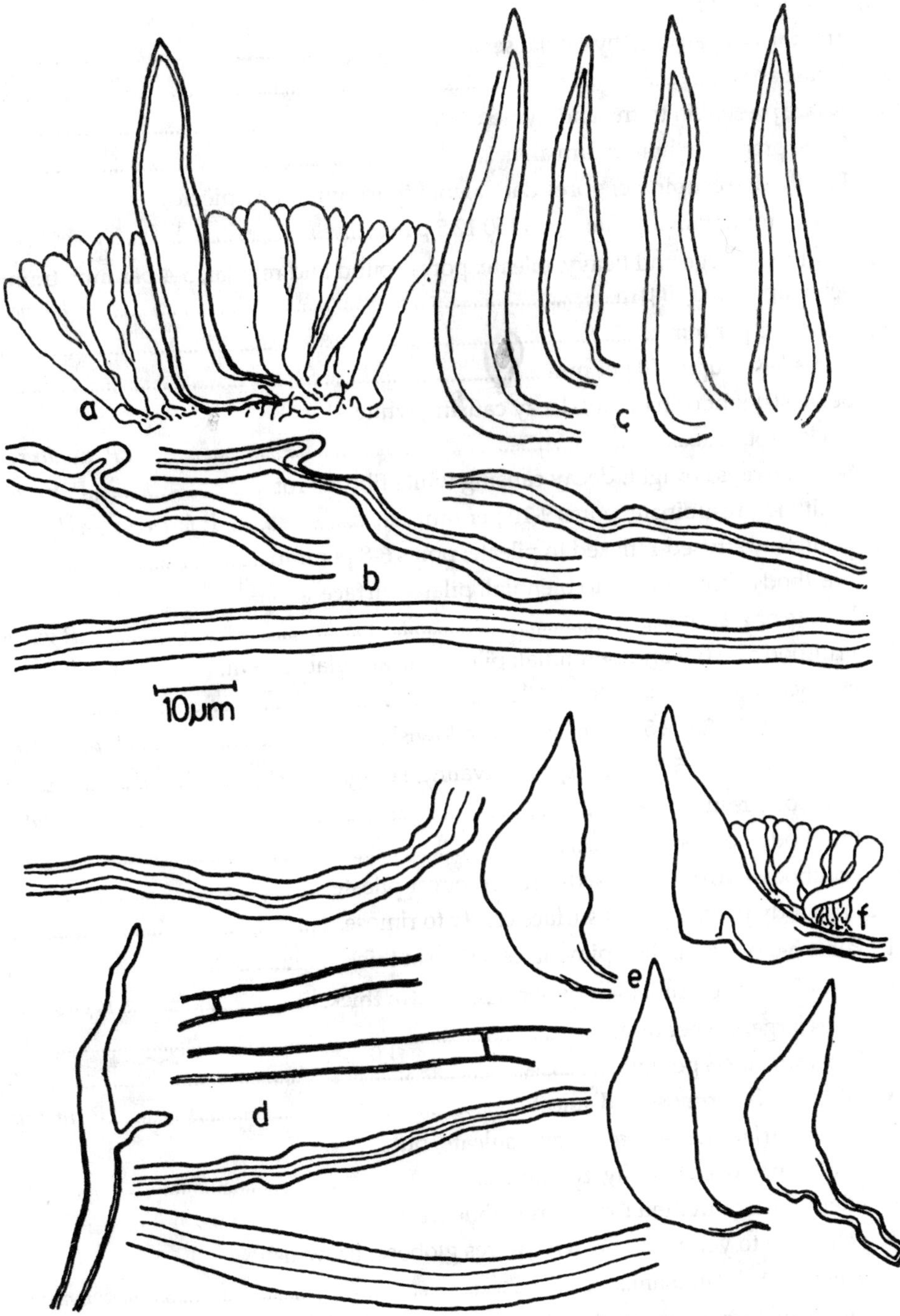

Fig. 8 : a - c : *Cyclomyces setiporus* : a - Setae with sterile hymenial elements; b - generative hyphae; c - setae.
d - f *Cyclomycees tabacinus* : d - generative hyphae; e - setate; f - setae with sterile hymenial elements.

KEY TO SPECIES

1. Tramal setae and/or hymenial setae present... 2
1′. Setate absent.. 9
2. Setate present in trama and hymenium.. 3
2′. Setae present in hymenium alone.. 4
3. Fruitbody resupinate; pores daedaleoid to irregular or irpicoid, 1-2per mm; tramal setate up to 150 12.5 µm... *P. contiguus*
3′. Fruitbody large and heavy, pileate; pores round and regular, 3-4 per mm; tramal setae up to 400 10 µm.. *P. hohnelii*
4. Pores 3-5 per mm...5
4′. Pores 6-8 or up to 10 per mm.. 6
5. Setae straight or hamate; decay causing white pocket rot... *P. wahlbergii*
5′. Setae always straight; decay causing white fibrous rot.............................. *P. ferreus*
6. Fruitbody resupinate; pores 9-10 per mm.............................. *P. ferrugineo-velutinus*
6′. Fruitbody effused-reflexed to pileate; pores 6-8 per mm.. 7
7. Fruitbody thin, annual to biennial; pileus surface hirsute or glabrous.. *P. gilvus*
7′. Fruitbody heavy and perennial; pileus surface glabrescent, crusty, rimose and sulcate... 8
8. Setae 12.5-15(20) × 5-7 µm; spores yellowish, globose.............................. *P. dependens*
8′. Setae (15-25(30) 5-7.5 µm; spores hyaline, sub-globose to ovoid................ *P. senex*
9. Fruitbody resupinate... *P. punctatus*
9′. Fruitbody not resupinate...10
10. Fruitbody narrowly effused-reflexed never pileate.. 11
10′. Fruitbody pileate; pileus surface crusty to rimose.. 12
11. Pores 5-6 per mm; dissepiments 50-100 µm thick.. *P. allardii*
11′. Pores 7-8(9) per mm; dissepiments 30-50 µm thick..................................... *P. gilvoides*
12. Pores large, 4-6 per mm... 13
12′. Pores small, 7-8 per mm... 15
13. Pileus surface rimose and sulcate... *P. rimosus*
13′. Pileus surface never rimose and sulcate, but crusty and concentrically, zonate... 14
14. Pileus surface matted tomentose while fresh, yellowish to yellowish brown; spores globose, brownish yellow, 4-4. 5 µm diam.. *P. adamantinus*
14′. Pileus surface glabrous, dark brown to black, spores sub-globose, often with an apex, yellowish brown, 5-6 × 4.5-5µm.. *P. sublinteus*

15. Fruitbody flabelliform to spathulate; pileus surface sepia brown to clay, velutinate while young, mostly persistent; pores 5-8 per mm; Spores 4-5.5 × 3-5 µm................................ *P. fastuosus*
15′. Fruitbody applanate to ungulate or triquetrous; pileus surface brownish to black, sulcate and rimose..16
16. Pileus surface dark brown to black; pores 7-9 per mm; context homogeneous, without black crusty lines.............................. *P. pseudosenex*
16′. Pileus surface light brown to sepia; context with black crusty lines.. 17
17. Fruitbody applanate to narrowly ungulate; pores 7-8 per mm; generative hyphae 1.5-2.5 µm, thin-walled; spores brownish yellow, globose, 4-5 µm in diam............................ *P. nilgheriensis*
17′. Fruitbody triquetrous; pores 8-10(12) per mm; generative hyphae thick-walled, 2.5-3 µm; spores yellowish to yellowish brown, sub-globose to globose, (3) 4-5 × (3)3.5-4 µm... *P. troyanus*

PHELLINUS ADAMANTINUS (Berk.) Ryv.
Norw. J. Bot. 19 : 234, 1972.

Fig. 9 a-c, Pl., Fig.

Polyporus admantinus Berk., Hook. J. Bot. 6 : 141, 1854.
Fomes adamantinus (Berk.) Sacc., Syll. Fung. 6 : 204, 1888.
Polyporus goethartii Bres., Ann. Mycol. 8 : 588, 1910.

Fruitbody perennial, solitary, epixylous, pileate, sessile, laterally attached with a broad base, semi-circular, woody hard, triquetrous; 8.5 × 4.5 × 2.5 cm. Pileus surface yellowish brown (5E8), older portions becoming rust brown (6E8) to dark grey (1F1); matted tomentose while fresh, becoming glabrascent, older portions heavily crusted, concentrically zonate; margin rigid smooth and rounded. Pore surface wax yellow (3B5), even and smooth; pores small, extending almost up to the margin, round to somewhat ovate, 5-6 per mm; pore mouth 75-125 µm broad; dissepiments 35-100(150) µm thick; pore tubes stratified, concolorous or slightly darker than the pore surface, 1.5-2 mm long. Context golden yellow (5B7) with sun yellow (2A5) strand, darker in KOH, 1-2 mm thick, fibrous, consisting of compact, parallel hyphae.

Hyphal system dimitic. Generative hyphae hyaline to sulphur yellow, thin-walled, sparsely branched, simple-septate, 2.5-3 µm in diam. Skeletal hyphae brownish yellow, thick-walled with broad and distinct lumen, unbranched, infrequently simple-septate, 4.5-5.5 µm in diam. Setate none. Basidia not observed. Basidiospores brownish yellow, globose, thick-walled, smooth, nonamyloid; 4-4.5 µm in diam.

DECAY : White rot.

SPECIMEN EXAMINED : On unidentified decaying tree trunks lying on the forest floor, Kaduvakkunnu, Nadugani forests, Nilambur Forest Division, Malappuram Dist., Herb. CALI No. G64, 28-2-1984.

GEOGRAPHICAL DISTRIBUTION : India, Indonesia, New Guinea.

REMARKS : Bakshi (1971) described the East Himalayan collections of this polypore as *Fomes adamantinus* and his collections have markedly thicker context and longer pore tubes than those of the present collection. The tomentose to glabrescent, crusty, azonate pileus surface, yellowish context, lack of setae, and small, globose, thick-walled spores are characteristic of this species.

PHELLINUS ALLARDII (Bres.) Ryv.

Norw. J. Bot. 19 : 234, 1972. Fig. 9. d-e, Pl., Fig.

Fomes allardii Bres, Bull. Jard. Bot. Bruxelles 4 : 19, 1911.

Fruitbody perennial, resupinate to somewhat effused-reflexed, firmly adhering to the substrate, margin in some regions slightly reflexed, tough and woody in consistency, spreading on the lower surface of tree trunks in sheets of 10 × 10 cm or more, -0.6 cm thick. Pileus surface narrow, black, glabrous and slightly crusty. Pore surface champagne (4B4) almost clay (5D5) when dry, smooth, highly uneven; pores extending to the margin, round to ovate, regular, small and not visible to naked eye, 5-6 per mm; pore mouth 90-125 μm wide; dissepiments 50-75 (100) μm thick; pore tubes stratose, concolorous with the pore surface, older tubes whitish inside and filled with context-tissue, 1.5-2 mm long in each stratum, growth often uneven, oblique or straight. Context a very thin layer in between pore tubes, concolorous with the pore tubes, consisting of compactly arranged parallel hyphae.

Hyphal system dimitic. Generative hyphae hyaline, thin-walled to slightly thick-walled, sparsely branched, simple-septate, 2-3 μm in diam. Skeletal hyphae slightly brownish yellow, thick-walled with distinct lumen, usually unbranched or with short branches near extremities, non-septate, straight and equal in width, often flexuous and prominently twisted at ends, 2-4 μm in diam. Setae none. Basidia and basidiospores not observed.

DECAY : White stringy rot, white mycelial tufts commonly occur in between wood crevices.

SPECIMEN EXAMINED : On living *Terminalia alata* Heyne ex Roth. (COMBRETACEAE), Kamathalamudi, Tunacadavu, Parambiculam, Teak Plantation Division, Palghat Dist., Herb. CALI No. G215, 12-9-1985.

GEOGRAPHICAL DISTRIBUTION : India, Pakistan, North America, Mexico, Cuba, Costa Rica, East Africa.

REMARKS : Bakshi (1971) reported *P. allardii* from the West Himalayas, as causing decay in living oak (Quercus spp.) The present collection agrees well with the description given by Bakshi (1971). The North American (Lowe, 1957) and East African (Ryvarden & Johansen, 1980) collections have smaller pores.

PHELLINUS CONTIGUUS (Pers. ex. Fr.) Pat.

Eassi Tax. P. 97, 1900. Fig. 10. a-d; Pl., Fig.

Boletus contiguus Pers., Syn. Meth. Fung. P. 544, 1801)

Polyporous contiguus (Pers.) ex Fr., Syst. Mycol. 1 : 378. 1821.

P. cribrosus Pers., Mycol. Eur. 2 : 96, 1825.

P. racodicodes Pers., Ibid. 2 : 113, 1825.

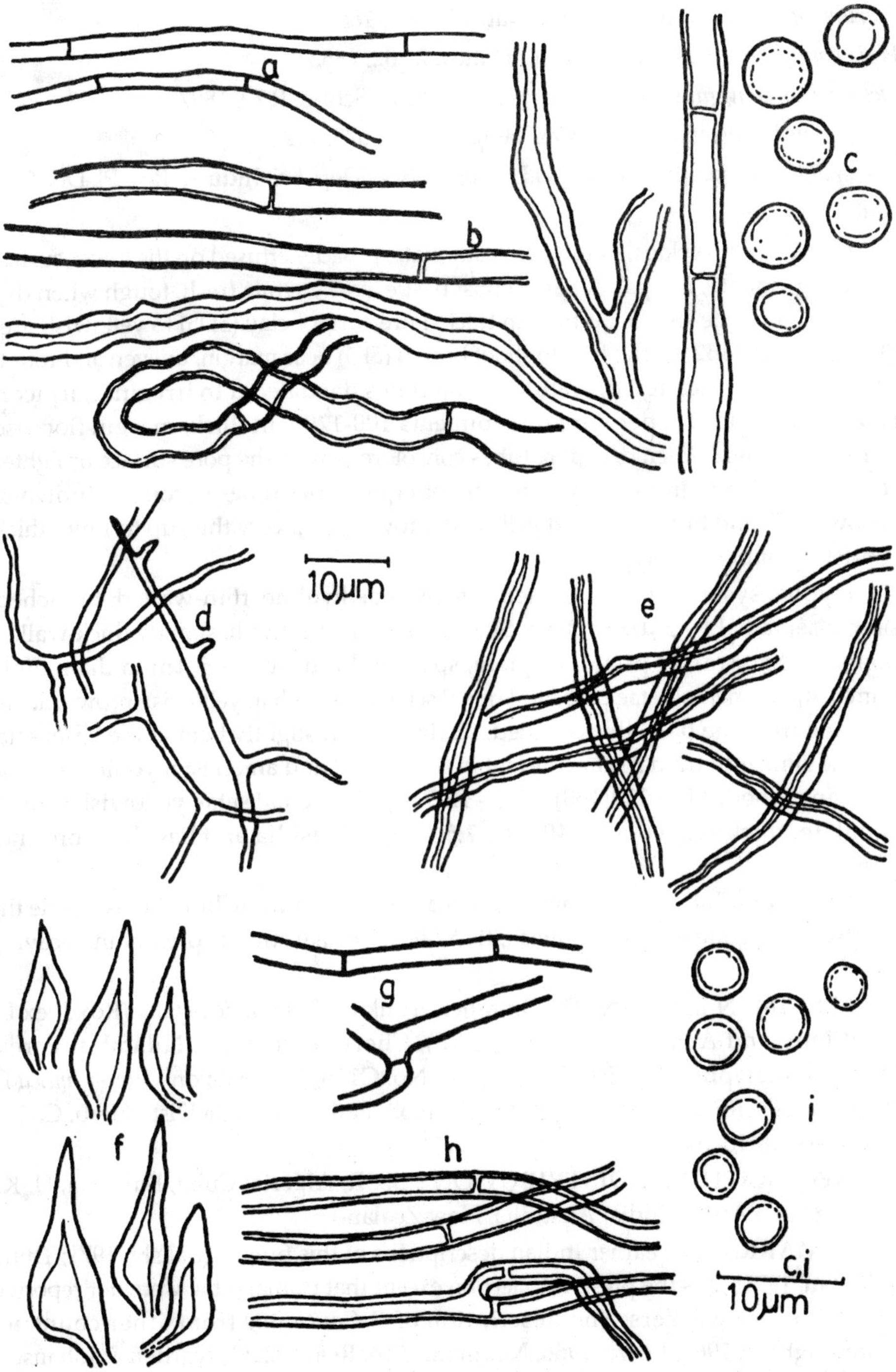

Fig. 9 : a-c : *Phellinus adamantinus* : a - generative hyphae; b - skeletal hyphae; c = basidiospores.
d - e *Phellinus allardii* : d - generative hyphae; e - skeletal hyphae.
f - i *Phellinus dependens* : f - setae; g - generative hyphae; h - skeletal hyphae; i - basidiospores.

P. crypthacanthus Mont., Ann. Sci. Nat. 4,5 : 369, 1857.

Poria contiguus (Pers. ex Fr.) Karst., Hattav. 2 : 82, 1882.

Ochroporus contiguus (Pers. ex Fr.) Schroet., Pilze Schles. P. 79, 1897.

Polyporous holubyanus Velen., Ces. Honby, 4-5 p. 673, 1922.

Fuscoporia contigua (Pers. ex Fr.) G.H. Cunn., N.Z. Dep. Sci. Industr. Res. Pl. Dis. Div. Bull. 73 : 4, 1948.

Fruitbody epixylous, perennial, resupinate, broadly effused on the lower surface of the log, adnate, inseparable from host tissue, corky while fresh, tough when dry, spreading in large sheets, up to 1 cm thick. Pore surface clay (5D5) to yellowish grey (3D2) or putty (4B2), corn (4B5) to shoot brown (5F5) near margin, uneven and rough; pores irregular, scattered, angular, sometimes daedaleoid to irregular, irpicoid towards margin, 1-2 per mm; dissepiments 100-120 µm thick; margin floccose, sterile, hairs rigid when dry; pore tubes concolorous with the pore surface or lighter, stratified, 2-3 mm in each layer, mostly oblique, older tubes becoming brownish yellow (5C7) and brittle. Context yellowish brown (5E8), very thin, up to 1 mm thick, consisting of parallel hyphae.

Hyphal system dimitic. Generative hyphae hyaline, thin-walled, branched, simple-septate, 1.5 - 2 µm in diam. Skeletal hyphae yellowish, slightly thick-walled, non-septate or rarely pseudo-septate, sparsely branched, 3-5 µm in diam. Both tramal and hymenial setae present. Tramal setae yellowish to yellowish brown, acute, tapering from the base, thick-walled, straight, often slightly bent at the base; setal hyphae similar to tramal setae present in the context and amongst mycelium seen in decaying wood, 115 - 130(150) × 7.5 - 12.5 µm; hymenial setae yellowish brown, subulate, thick-walled, (25) 40-50 × 7.5 - 9 µm. Basidia and basidiospores not observed.

DECAY : White rot, bark and sapwood becoming a white brittle mass while the heartwood is quite tough. yellowish, floccose, mycellial strands present in decaying wood.

SPECIMEN EXAMINED : On fallen trunk of *Terminalia arjuna* (Roxd. ex D. C.) Wight and Arn (COMBRETACEAE), Cherupuzha river, Nilambur Forest Division, Malappuram Dist., Herb. CALI No. G155, 7-3-1985; on *Vateria indica* L. (DIPTEROCARPACEAE), Aramba, Achenkoil, Pathanamthitta Dist., Herb. CALI . No. G364, 18-10-1986.

GEOGRAPHICAL DISTRIBUTION : USSR, Algeria, Cuba, Bulgaria, U. K., Jamaica, East Africa, India, Australia, New Zealand.

REMARKS : The earlier Indian descriptions of this fungus (Bakshi, 1971; Thind & Rattan, 1971b) fits the present collection except that tramal setae were not reported by the former workers. The description of *P. Contiguus* from other countries (Cunningham, 1965; Lowe, 1966; Neuman, 1914; Rea, 1922; Ryvarden & Johansen, 1980) agree with the present collection except for minor differences in some cases. The African collections (Ryvarden & Johansen, 1980) have smaller pores and slightly broader generative hyphae.

Persoon (1801) placed this fungus among pileate species but the truely resupinate nature of the fruitbody induced several others (Bakshi, 1971; Lloyd, 1898-1925; Lowe, 1966; Rea, 1922) to include it in the genus *Poria* Pers. ex Gray while Cunningham (1965) accommodated it in *Fuscoporia* Murr.

PHELLINUS DEPENDENS (Murr.) Ryv.

Norw. J. Bot. 19 : 234, 1972. Fig. 9. f-I; Pl. Fig.

Pyropolyporus dependens Murr., North Am. Fl. 9 : 106, 1908.

Fomes dependens (Murr.) Sacc. & Trott., Syll. Fung. 21 : 292. 1912.

Fulvifomes dependens (Murr.) Murr., Trop. Polyp. P. 88, 1915.

Fruitbody perennial, solitary to imbricate, effused-reflexed initially, later becoming pileate, attached with a broad lateral base; semi-circular to irregular, broadly applanate to somewhat triquetrous or ungulate, tough and woody; 2-7 x 1.5-3.5 x 1.5-3cm. Pileus surface earth coloured (5F2) to black, glabrous when young and old, rough and uneven, concentrically ridged, ridges sharp, sulcate, faintly crusty, sometimes radially cracking with age; margin rigid, wavy or entire. Pore surface olive brown (4D8) to yellowish brown (6F7) when old, growth not uniform, smooth and mostly uneven, margin slightly projecting over previous year's growth; pores minute, not visible to naked eye, extending to the margin, round to angular, regular, 6-8 per mm; pore mouth 85-110 μm broad; dissepiments 40-60 μm thick; pore tubes stratified, golden brown (5D7), 1.5-2.5 mm long in each stratum, old tubes darker with white insides or undifferentiated, invaded by context tissue. Context concolorous with pore tubes, absent between pore layers, about 2mm thick, upper region agglutinated to form a black crust about 1 mm thick, lower portion fibrous, homogeneous, compact.

Hyphal system dimitic. Generative hyphae sparse, hyaline, thin to slightly thick-walled, branched, simple-septate, 2.5-3.5 μm in diam. Skeletal hyphae yellow to brownish yellow, thick-walled with broad lumen, simple-septate, branched, straight and equal in width, 3.75-4.5 μm in diam. Hymenial setae present but not abundant, yellowish brown, rusty brown in KOH, ventricose, acuminate, thick-walled, straight, 12.5-15(20) × 5-7 μm. Basidia not observed. Basidiospores light yellow, golden brown in KOH, globose to subglobose, thick-walled, smooth, nonamyloid, 4-5 μm in diam.

DECAY : White pocket rot of heart wood. The sapwood and bark region intact except for the portion just beneath the sporophore. Sporophores occur at the basal region of the living trunk up to 2 m.

SPECIMEN EXAMINED : On living trunk of *Terminalia paniculata* Roth (COMBRETACEAE), near Perumpara (W), Malakkapara, Sholayar forests; Trichur Dist., Herb. CALI No. G350, 30-8-1986; on *Mallotus philippensis* (Lamk.) Muel-Arg. (EUPHORBIACEAE), Kathiyur Reserve Forests, Wyanaad Division, Wyanaad Dist., Herb. FRI No. 4335, K. Bagchee, 23-1-1946.

GEOGRAPHICAL DISTRIBUTION : North, Central and South America, India.

REMARKS : Bakshi's (1971) description of this fungus as *Fomes dependens*, based on a collection from Kerala by Bagchee in 1946, shows spores bigger than those of the present collection. A common species of American tropics, this fungus was described in detail by many American workers (Lowe, 1952, 1957; Murrill, 1915; Overholts, 1953) but the presence of broadly ellipsoid spores seems unique to American specimens. Ryvarden & Johansen (1980) described this species with subglobose spores similar to the present collection.

PHELLINUS FASTUOSUS (Lev.) Ryv.

Norw. J. Bot. 19 : 234, 1972. Fig. 10. e-g; Pl., Fig.

Polyporus fastuosus Lev., Ann. Sci. Nat. 3, 2 : 190, 1844.

P. spadiceus Berk., Ann. Nat. Hist. 3 : 338, 1838, nomen nudum, non Jungh., 1838.

Fruitbody epixylous, perennial, pileate, sessile, attached with a broad lateral base, solitary, sometimes imbricate and confluent, easily detachable from the host, flabelliform to spathulate, applanate to conchate, woody hard, 8-22.5 × 6-17.5 × 1.5.5 cm. Pileus surface sepia brown (5F4), oak brown (5D6), clay (5D3), sallow (4D3) to dark brown (6F8) or black, colour varying from one fruitbody to another,, finely velutinate while young soon becoming glabrous, concentrically striate to sulcate, crusty, sometimes warty, often lobate towards margin or deeply incised; margin entire, smooth, finely velutinate, thin, rounded. Pore surface yellowish brown (5D8), linoleum brown (5E7), mustard brown (5E6), honey yellow (4D6), becoming lighter towards margin, colour often varying according to season and locality, rough and uneven, growth not uniform; pores often in patches, extending almost up to the margin, small, round, regular, (5) 6-8 per mm, pore mouth 75-95(120) µm wide; dissepiments 50-75 µm thick; pore tubes stratified, concolourous or slightly darker than the pore surface, fresh layer 1-2 mm long, old tubes up to 3mm long, straight, old tubes filled with context-tissue. Context putty (4B2) to yellowish brown (5F8), absent between pore layers, fibrous, homogeneous, sometimes with a thin black crusty line about 2 mm below the upper crust, 3-8 mm thick, of almost parallel hyphae compactly arranged.

Hyphal system dimitic. Generative hyphae hyaline, thin to slightly thick-walled, branched, simple-septate, less represented in context, 2-3 (4.5) µm in diam. Skeletal hyphae yellow to yellowish brown, thick-walled with broad lumen, usually unbranched or with small apical branches, frequently simple-septate, 2.5-6 µm in diam. Setae none. Basidia not observed. Basidiospores subglobose to globose, yellow, reddish brown to rusty brown in KOH, thick-walled, smooth, 4-5 µm in diam. when globose, otherwise (3.5)4-4.75(5.5)× (3)4-5 µm.

DECAY : A white pocket rot; the region just beneath the sporophore appear brownish.

SPECIMEN EXAMINED : On trunks of living *Swietenia macrophylla* King (MELIACEAE), Divisional Forest Office Campus, Nilambur, Malappuram Dist., Herb. CALI No. G154, 7-3-1985; Forest Range Office premises, Karualai, Nilambur, Malappuram Dist., Herb. CALI No. G163, 17-3-1985; on *S. mahagony (Linn.)* Jacq. Panangode Block, Nilambur, Malappuram Dist., Herb. FRI No. 4334, K.Bagchee, 19-1-1946; on trunk base of living *Xylia xylocarpa* (Roxb.) Theob. (MIMOSACEAE), KFRI sub-section campus, Nilambur, Malappuram Dist., Herb. CALI No. G40, 16-9-1983; on dead *Ficus sp.* (MORACEAE), Vengoli peak, Tunacadavu,

Parambiculam, Palghat Dist., Herb. CALI No. G214, 11-7-1985; on trunks of living *Schleichera oleosa* (Lour.) Oken (SAPINDACEAE), Forest Range Office premises, Karulai, Nilambur, Malappuram Dist., Herb. CALI No. G161, 7-3-1985; on fallen trunk of *Grewia tiliifolia* Vahl (TELIACEAE), Kamathalamudi, Tunacadavu, Parambiculam, Palghat Dist., Herb. CALI No. G218, 12-7-1985; on unidentified decaying tree trunks, Parambiculam Palghat Dist., Herb. CALI No. G274, 16-1-1986; Forest Range Office premises, Karulai, Nilambur, Malappuram Dist., Herb. CALI No. G164, 7-3-1985.

GEOGRAPHICAL DISTRIBUTION : North Amercia, Mexico, Yucatan, East Africa, Singapore, Philippines, India, Pakistan.

REMARKS : *P. fastuosus* which has a fairly wide host range cause severe heart rot of *Schleichera oleosa* and *Swietenia spp.* in Nilambur Forests. This fungus is common in man-made gardens and disturbed forests but rare in evergreen forests. The sporophores are mainly confined to the basal part of the tree trunk but rarely they are seen up to 10 m height.

PHELLINUS FERREUS (Pers.) Bourd. & Galz.
Hymen. Fr. P. 627, 1928. Fig. 11. a-c; Pl., Fig.

Polyporus ferreus Pers., Mycol. Eur. 2 : 89, 1825.
Mucronoporus fulvidus Ell. & Ev., Proc. Phil. Acad. Sci. P. 323, 1894.
Poria usambarensis Henn., Engl. Bot. Jahrb. 38 : 108, 1905.
Fuscoporia fulvidia (Ell. & Ev.) Murr, North Am. Fl. 9 : 5, 1907.
Poria cylindrispora Lloyd, Mycol. Letter 65 : 9, 1917.
P. ferrea (Pers.) Overh.,Mycologia 23 : 117, 1931.
Fuscoporia ferrea (Pers.) G.H. Cunn., N. Z. Dep. Sci. Industr. Res. P1. Dis. Div. Bull. 73 : 7, 1948.

Fruitbody annual to biennial, resupinate, spreading on the lower surface of the log, adnate, 2-3 mm thick, thickness uniform, woody hard, spreading in sheets of 10 × 5 cm or more area in some. Pore surface golden (4C6) to amber yellow (4B6), reddish yellow (4A8) to reddish brown (8E8) towards periphery, rough, uneven, growth not uniform, pores small, visible to naked eye, oblique, extending to the margin, round to angular, not regular, 3-4(5) per mm; pore mouth 140-175 µm wide; dissepiments 80-100 µm thick; pore tubes concolorous with the pore surface, stratified, 1-2 mm long in each stratum, tube insides slightly greyish. Context a thin layer above pore tubes. absent between pore layers, yellow (2A8) to Naples yellow (3B7), less then 0. 5 mm thick, of compact radiating hyphae.

Hyphal system dimitic. Generative hyphae, thin-walled, well branched, simple-septate, 1. 5-2. 5 µm in diam. Skeletal hyphae yellowish, thick-walled with broad and distinct lumen, unbranched, non-septate, 3-4.5 µm in diam. Setae present in hymenium, brownish yellow, subulate to lanceolate, straight, thick-walled, 40-50 × 6-8. 5(10) µm Basidia and basidiospores not observed.

DECAY : Causing white rot; yellowish, thick-walled hyphae similar to that of context forming a mat just beneath the bark. Both sapwood and heartwood become a white fibrous to brittle mass.

SPECIMEN EXAMINED : On fallen branches of *Vateria indica* L. (DIPTEROCARPACEAE), in a semi-deciduous forest, Aramba, Achankoil, Kottayam Dist., Herb. CALI No. G365, 18-10-1986.

GEOGRAPHICAL DISTRIBUTION : New Zealand, Australia, North America, India.

REMARKS : The present collection agrees well with the description of *Phellinus ferreus* by Revarden & Johansen (1980). While Cunningham (1965) reported smaller setae, Gilbertson (1979) reported smaller setae and spores for this fungus. This is the first record of its occurrence in India.

PHELLINUS FERRUGINEO-VELUTINUS (Henn.) Ryv.

Norw. J. Bot. 19 : 234, 1972. Fig. 10. h-l; Pl., Fig.

Poria ferrugineo-velutinus Henn., Hedwigia 44 : 59, 1905.

Fomitopsis flavomarginata Murr., North Am. Fl. 9 : 1, 1907.

Fruitbody perennial, broadly effused, spreading on the lower surface of a fallen tree trunk, adnate; inseparable from the host, hard woody while fresh, slightly brittle on dying, margin narrow to broad, adpressed, spreading in sheets of 10 × 10 cm or more in area, up to 1 cm thick. Pore surface bronze brown (5E5) while fresh, sepia brown (5F4) towards margin, chamois (4C5) when dry; smooth and even; pores minute, not visible to naked eye, absent towards margin, round, regular, 9-10 per mm; pore mouth 75-90(100) μm wide; dissepiments 25-35(50) μm thick; pore tubes yellowish brown (5D8, 5E8), stratified, each stratum 2-3 mm, oblique or straight. Context a very thin layer in between pore layers, concolorous with pore tubes, less than 0.5mm thick, homogeneus, of compact parallel hyphae.

Hyphal system dimitic. Generative hyphae hyaline, thin-walled, simple-septate, 2-4 μm in diam, occasional in context. Skeletal hyphae yellowish brown, thick-walled with a narrow and distinct lumen, usually unbranched or with short branches towards extremities, sparsely simple-septate, cylindrical and flexuous, 2.75-4 μm in diam. Setae present in hymenium, yellowish brown, ventricose, thick-walled, scattered, 12-21 × 4.5-7.5 μm. Basidia clavate, short, 2 or 4-spored, 6-8 × 3.5-4.5 μm; sterigmata up to 2 μm long. Basidiospors hyaline, ovoid to sub-globose, thin-walled, smooth, nonamyloid, often with 1 or 2 oil droplets, 2.5-3 × 2.5 μm

DECAY : White rot.

SPECIMEN EXAMINED : On fallen branches of unidentified trees in a semi-deciduous forest, Mundakai, near Meppadi, Wyanaad Dist., Herb. CALI No. G249, 12-9-1985.

GEOGRAPHICAL DISTRIBUTION : Tropical America (Brazil, Cuba, Southern United States), India.

REMARKS : Out of the two closely related species, *Phellinus johnsonianus* (Murr.) Ryv. differs in having longer and wider setae, while *P. glaucacens* (Petch) Ryv., a Sri Lankan species, has slightly longer and wider setae and bigger spores. This is the first report of this fungus from India.

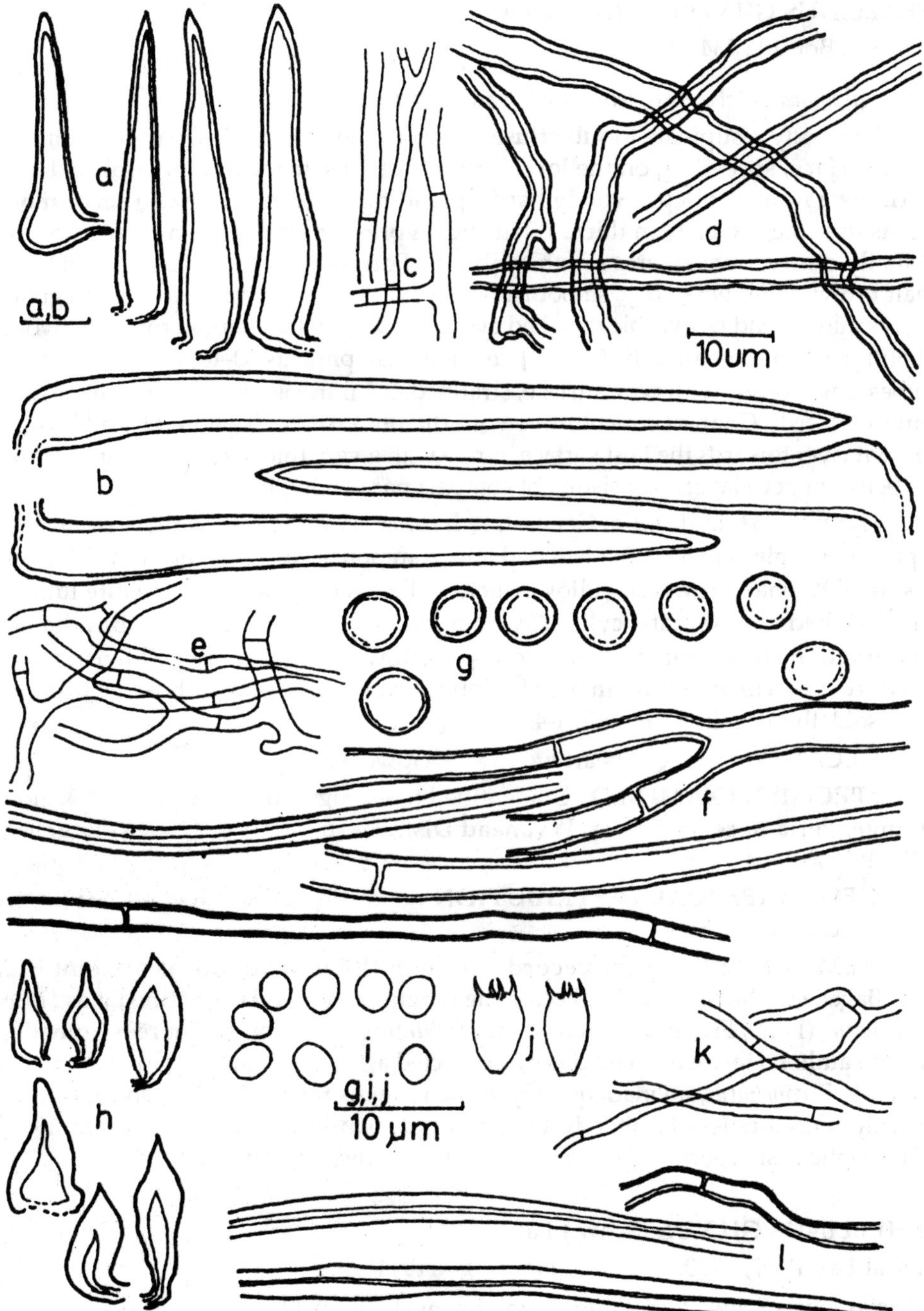

Fig. 10 : a-d *Phellinus contiquus* : a-hymenial setae; b-tramal setae; c-generative hyphae; d-skeletal hyphae.

e-g *Phellinus fastuosus* : e-generative hyphae; f - skeletal hyphae; g-basidiospores.

h-l *Phellinus ferrugineo-velutinus* : h-setae; i-basidiospores; j-basidia; k-generative hyphae; l-skeletal hyphae.

PHELLINUS GILVOIDES (Petch) Ryv.

Norw. J. Bot. 19 : 234, 1972. Fig. 11. d-f; Pl., Fig.

Poria gilvoidea Petch, Ann. Roy. Bot. Grad. Peradeniya 6 : 138, 1916.

Fruitbody resupinate to sub-effused-reflexed, adnate and inseparable from the host surface, spreading on the lower surface of logs, margin rarely uplifted and forming a narrow pileus, woody hard, spreading in sheets of 10 × 10 cm or more areas in some, up to 1 cm thick. Pileus when present, dark brown (8F6) to black, usually crusty, uneven and rough, glabrous. Pore surface yellowish clay (5D5) to hair brown (5E4) when dry, smooth, cracking laterally with age, wavy to irregular; pores minute and not visible to naked eye, extending to the margin, round to ovoid, 7-8(9) per mm; pore mouth 100-120 µm wide, dissepiments 30-40(5) µm thick; pore tubes concolorous with pore surface, stratified, 2-3 mm long in each stratum, inside tubes whitish. Context concolorous with the pore tubes; delimited by a blackish brown layer towards the host surface, rarely with a very thin layer (less than 0.5 mm) in between pore layers, consisting of compact parallel hyphae.

Hyphal system dimitic. Generative hyphae hyaline, thin-walled, branched, sparsely simple-septate, 1.5-2 µm in diam., common in dissepiments and old tubes. Skeletal hyphae brownish yellow, thick-walled with broad and definite lumen, unbranched, non septate, cylindrical, 3.5-5 µm in diam. Setae none. Basidia not observed. Basidiospores hyaline to pale yellow, slightly more yellowish when mature, brownish yellow in KOH, globose to slightly sub-globose or broadly ellipsoid, thick-walled, smooth, 4-4.5 × 3-3.5 µm.

DECAY : White rot of heart wood and sapwood.

SPECIMEN EXAMINED : Unidentified forest logs, host not identified, Kanoth Range, Peria Reserve Forest, Wyanaad Dist., Herb. FRI No. 4253, K. Bagchee, 27-11-1946.

GEOGRAPHICAL DISTRIBUTION : A rarely known species; Sri Lanka (Type Locality), India.

REMARKS : From the records of the FRI herbarium, it is evident that Dr. Bagchee, who originally collected the fungus from Kerala, identified it as *Fomes conchatus* (Pers.) Fr. but it proved to be *Phellinus gilvoides* on recent re̅examination by the author. However it could not be collected again from Kerala during the recent survey. *F. conchatus*, an inadequately known fungus from India (Bakshi, 1971), has a truly effused-reflexed fruitbody, big pores and abundant setae (Cunningham, 1965). This is the first report on the occurrence of *P. gilvoides* in India.

PHELLINUS GILVUS (Schw.) Pat.

Essai Tax. P. 97, 1900. Fig. 11. g-k; Pl., Fig.

Boletus gilvus Schw., Syn. Fung. Carol. Super. 1 : 96, 1822.

Polyporus gilvus (Schw.) Fr., Elench. Fung. P. 104, 1828.

P. scruposus Fr., Epicrisis p. 473, 1838.

P. rubiguosus Berk., Ann. Nat. Hist. 3 : 324, 1839.

P. lichnoides Berk., Ibid. 2, 13 : 204, 1840.

P. isidiodes Berk., Hook. Lond. J. Bot. 2 : 515, 1843.

P. spurcus Lev., Ann. Sci. Nat. 3, 5 : 135, 1846.

P. holosclerus Berk., Lond. J. Bot. 6 : 501, 1847.

P. breviporus Cke., Grevillea 12 : 17, 1847.

Trametes pertusa Fr., Kungl. Vetensk, Akad, Handb. p. 130, 1848.

Polystictus licnoides (Mont.) Fr., Nov. Symb. P. 76, 1851.

P. breviporus (Cke.) Sacc., Syll. Fung. 6 : 287, 1888.

Placodes Fucatus Quel., Fl. Myc. P. 399, 1888.

Polyporus illicola Henn., Engl. Bot. Jahrb. 32 : 39, 1902.

Haplopilus gilvus (Schw.) Murr., Bull. Torrey Bot. Club 31 : 418, 1904.

H. licnoides (Berk.) Murr., North. Am. Fl. 9 : 81, 1908.

Fomes gilvus (Schw.) Lloyd, Mycol. Letter 42 : 6, 1912.

Polyporus hookeri Berk. ex Lloyd, Mycol. Writ. 4 : 348, 1915.

Fomes taxodii Murr., Bull. Torrey Bot. Club 65 : 661, 1939.

F. scruposus (Fr.) G.H. Cunn., N.Z. Dep. Sci. Industr. Res. Pl. Dis. Div. Bull. 79 : 11, 1948.

Phellinus scruposus (Fr.) G.H. Cunn., N.Z. Dep. Sci. industr. Res. Bull. 164 : 230, 1965.

Fruitbody epixylous, annual or reviving during the second season, pileate, sometimes effused-reflexed, sessile, solitary or imbricate, often confluent, attached with a broad lateral base, semi-circular to dimidiate, coriaceous to corky while fresh, woody when dry; 2.5-10 × 2-7 × 0.3-0.7 cm. Pileus surface brown (6E6), yellowish brown to brownish grey (7D2) towards margin, with concentric bands of raw Sienna (6D7), reddish brown (8E8) or dark brick red (7D7), colours fading in irregular patches or uniform, glabrous or rough with agglutinated hyphae, the latter more prominent towards the base, certain sporophores radially wrinkled; margin thicker, rounded, regular to wavy or lobed. Pore surface yellowish brown (5E5) while young, reddish brown (8D6) when old, margin golden yellow (5B7) to lemon yellow (3B8), velvety in appearance while fresh, even and smooth; pores minute, absent near margin, angular to oval to round, 6-8 per mm; pore mouth 60-90 µm wide; dissepiments 35-65(85) µm thick; pore tubes concolorous with the pore surface, sometimes stratified, straight, up to 5 mm long. Context golden brown (5E8) to golden yellow (5E7), reddish brown (8E6) towards the crust, homogeneous, shiny, 2-3 mm thick, of parallel hyphae.

Hyphal system dimitic. Generative hyphae hyaline, thin-walled, branched simple septate, 2.5-3 µm in diam. Skeletal hyphae brownish yellow, thick-walled with narrow lumen, usually unbranched, 3.75-5.5 µm in diam. Setae present in hymenium, yellowish brown, thick-walled, ventricose to subulate, actute, 12.5-30 × 5-6.6 µm long. Basidiospores hyaline, sub-globose to ellipsoid, thin-walled, smooth, 3.5-4.5 × 3-4 µm.

DECAY : White rot.

SPECIMEN EXAMINED : Dead branches of *Terminalia bellerica* (Gaertn.) Roxb. (COMBRETACEAE), Wyanad Dist., Kuppadi, Wyanaad Dist., Herb. CALI No. G292, 20-6-1986; on decaying stumps of *Dendrocalamus strictus* (Roxb.) Nees (POACEAE), Kamathalamudi, Parambiculam, Palghat Dist., Herb. CALI No. G217, 12-11-19987; on decaying trunks of *Vateria indica* Linn., (DEPTEROCARPACEAE), Kanoth Range, Peria Reserve Forest, Wyanaad Dist., Herb FRI No. 4249, K. Bagchee, 24-1-1946; on decaying stumps of *Toona ciliata* Roem. (MILIACEAE) Kanoth Range, Peria Reserve Forest, Wyanaad Dist., Herb. FRI No. 4279, K. Bagchee, 25-1-1946; on decalying poles of *Tectona grandis* Linn. f. (VERBENACEAE), Vazhachal, Sholayar forests, Trichur Dist., Herb. CALI No. G81, 26-7-1984; Divisional Forest Office Campus, Nilambur Malappuram Dist., Herb. CALI No. G146, 7-3-1985; on unidentified logs and wood, Kaduvakkunnu, Nadugani forests, Nilambur, Malappuram Dist., Herb. CALI No. G49, 28-2-1984, Herb CALI No. G130, 13-11-1984; Botanical Garden, Calicut University campus, Malappuram Dist., Herb, CALI No. G111, 5-8-1983; Chooralmala, near Meppadi, Wyanaad Dist., Herb CALI No. G119, 21-9-1984; Mundakai, near Mepadi, Wyanaad Dist., Herb. CALI No. G245, 12-9-1985; Vengoli Peak, Parambiculam, Palghat Dist., Herb. CALI No. G209, 11-7-1985; Parambikulam forests, Palghat Dist., Herb. CALI No. G-220, 13-7-1985; Panthenthode, Silent Valley National Park, Palghat Dist., Herb. CALI No. G224, 15-9-1985.

GEOGRAPHICAL DISTRIBUTION : Cosmopolitan.

REMARKS : *P. gilvus* is of common occurrence in India (Bose, 1937a; Thind & Chatrath, 1957; Bakshi, 1971). It is often seen in the timber-depots of Kerala, causing white rot of commercially important woods.

PHELLINUS HOHNELII (Bres.) Ryv.
Prelim. Polyp. Fl. E. Africa p. 173, 1980.

Fig. 11. l-0; Pl., Fig.

Fomes hohnelii Bres., Ann. Mycol. 10 : 499, 1912.
F. magnosporus Lloyd, Mycol. Writ. 6 : 1014, 1920.

Fruitbody perennial, solitary, pileate, sessile, attached with a broad lateral base, semi-circular, sub-applanate to broadly ungulate, heavy, tough, woody; 27-57 × 17-30 × 7.5-15 cm. Pileus surface sepia brown (4F4) or smoke brown (4F2) to coal black (3F1), younger areas slightly wine yellow (3B3); rimose-crusty, concentrically sulcate, older parts breaking into irregular blocks, radial cracks more prominent, margin thicker, rounded, smooth. Pore surface yellowish clay (5D5) when fresh, becoming dark blonde (5D4) or nougat (5D3) or black when dry, somewhat uniform but rough, cracking irregularly with age; pores small, extending almost to the margin; pores round, regular, 3-4 per mm; pore mouth 150-190 µm broad; dissepiments 25-75(100) µm thick; pore tubes yellowish brown (5E8) becoming dark brown (6F8) when old, stratified, each layer about 0.6-1cm long, old tubes stuffed with context-tissue. Context yellowish brown (5E8) to dark brown (6F8), absent between pore layers, homogeneous, fibrous, up to 3 cm thick, of compact parallel hyphae.

Hyphal system dimitic. Generative hyphae hyaline thin-walled, branched, simple septate, 1.5-3 μm in diam. Skeletal hyphae brownish yellow, thick-walled with a distinct lumen, rarely branched at the extremities, simple-septate, 2.5-4 μm in diam. Both setal hyphae and hymenial setae present; setal hyphae brownish, thick-walled, acute, sometimes obliquely terminating in the hymenium, up to 400 μm long, 6-10μm broad; hymenial setae ventricose to acuminate, brownish, thick-walled, 18-27 × 6-9 μm. Basidia and basidiospores not observed.

DECAY : Associated with a uniform white rot of heartwood. Wefts of white mycelium is common in wood crevices. Collected from a moist deciduous stand. Several sporophores of *P. hohnelii* were found growing on a partially dead *Terminalia sp.*. fruitbodies arising on the trunk and main branches from as low as half-a-metre from the ground to almost 7m.

SPECIMEN EXAMINED : on living *Terminalia sp.* (COMBRETACEAE), Kamathalamudi, Tunacadavu, Teak Plantation Division, Parambiculam, Palghat Dist., Herb. CALI No. G219, 12-7-1985; G269, 12-11-1985.

GEOGRAPHICAL DISTRIBUTION : Philippines, India, Africa.

REMARKS : This is the first report on the occurrence of this polypore in India. *P. hohnelii* is close to *P. pachyphloeus* (Pat.) Pat., but the former differs in having sub-ungulate fruitbodies, larger pores, narrower setal hyphae and in the absence of setae on the mycelium present in decaying wood.

PHELLINUS NILGHERIENSIS (Mont.) G.H. Cunn.

N.Z. Dep. Sci. Industr. Res. Bull. 164 : 226, 1965.

Fig. 12. a-c; Pl., Fig.

Polyporus nilgheriensis Mont., Ann. Sci. Nat. 2, 18 : 1842.

Polystictus nilgheriensis (Mont.) Fr., Nova. Acta 3, 1 : 87, 1851.

Fruitbody annual to perennial, pileate and sessile, attached with a broad lateral base, imbricate, rarely confluent, semi-circular, applanate to narrowly ungulate, tough and woody, heavy; 11-15 × 4.5-7 × 0.5-1.5 cm. Pileus surface brownish grey (5E2) to sepia (5F4), often black, uneven and rough, glabrous, concentrically sulcate, crusty towards base, radially cracking with age; margin thick, rounded, rigid. Pore surface initially Naples yellow (3B7) soon becoming linoleum brown (5E7) or golden brown (5D7), yellowish brown (5E8) when dry, uneven and rough; pores minute, absent near margin, round, regular, 7-8 per mm; pore mouth 100-125 μm wide dissepiments 35-50 μm thick; pore tubes stratified, golden brown (5D7) to yellowish brown (5D8), each layer 2-3 mm long. Context golden brown (5D7), fibrous, 2-3 mm thick, often with a thin black line just beneath the upper crust, of compact parallel hyphae.

Hyphal system dimitic. Generative hyphae hyaline, thin-walled, less branched, sparsely simple-septate, infrequent in context, 1.5-2.5 μm in diam. Skeletal hyphae brownish yellow, thick-walled with broad and distinct lumen, unbranched but rarely branched near apices, central stem cylindrical, extremities flexuous, 3.75-5 μm in diam. Setae none. Basidia not observed. Basidiospores brownish yellow,

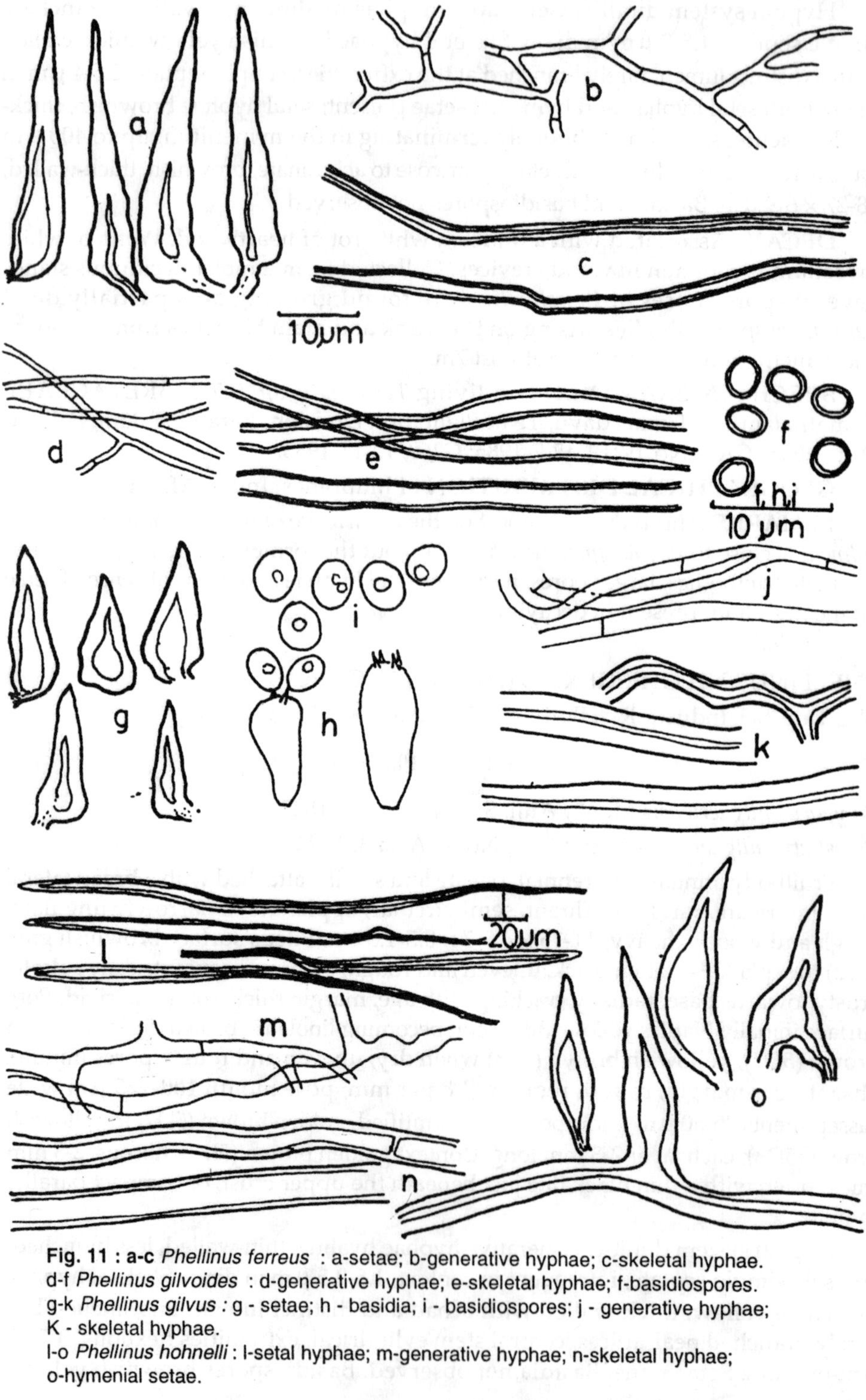

Fig. 11 : a-c *Phellinus ferreus* : a-setae; b-generative hyphae; c-skeletal hyphae.
d-f *Phellinus gilvoides* : d-generative hyphae; e-skeletal hyphae; f-basidiospores.
g-k *Phellinus gilvus* : g - setae; h - basidia; i - basidiospores; j - generative hyphae;
K - skeletal hyphae.
l-o *Phellinus hohnelli* : l-setal hyphae; m-generative hyphae; n-skeletal hyphae;
o-hymenial setae.

yellow brown to rusty brown in KOH, globose, rarely sub-globose, thick-walled, smooth 4-5 μm in diam.

DECAY : White rot

SPECIMEN EXAMINED : Dead, unidentified tree trunk, Kuppadi forests, near Sultan Battery, Wyanaad Dist., Herb. CALI No. G297, 20-6-1986.

GEOGRAPHICAL DISTRIBUTION : Cuba, Samoa, Tanzania, India.

REMARKS : Although the type of *P. nilgheriensis* was reported from India, (Bose, 1922a; Bakshi, 1971) a complete description of this species from India was lacking so far. *Phellinus nilgheriensis* from East Africa (Ryvarden & Johansen, 1980) has sub-globose spores while the spores are mostly globose in the present collection. It is noteworthy that the present collection was made close to the type locality.

PHELLINUS PSEUDOSENEX (Murr.) Bond. & Herr.

Fig. 12. d-f. Pl. Fig.

Pyropolypores pseudosenex Murr., North Am. Fl. 9 : 107, 1908.

Fomes pseudosenex (Murr.) Sacc. & Trott, Syll. Fung. 21 : 292, 1912.

Fruitbody perennial, pileate, sessile, imbricate and confluent, rarely solitary, conchate to irregular, laterally attached with a broad base, hard woody, 8-12(30) × 9-13(20) × 4-6(8) cm. Pileus surface dark brown (6F8) to black, often yellowish brown (5E5) at the margin, glabrous when mature, finely tomentose near the margin, heavily crusty, uneven, concentrically sulcate, radially splitting to irregular blocks when old. Pore surface olive yellow (3D7) to mustard brown (5E6), shiny, smooth but uneven, growth irregular; pores minute and not visible to naked eye, absent near margin, round, regular, 7-9 per mm; pore mouth 75-90 μm wide; dissepiments 35-50 μm thick; pore tubes stratified, yellowish brown (5E8) to rust brown (6E8), tubes 2-3 mm in each layer, oblique or straight. Context present in between pore layers, rust brown (6E8), bright with a silky lustre, often with a thin black line running irregularly, fibrous, 1-2 mm in each stratum, old tubes filled with context-tissue, context hyphae and tubes cannot be differentiated in older portions.

Hyphal system dimitic. Generative hyphae hyaline, thin to slightly thick-walled, branched, simple-septate, septations infrequent, often observed in dissepiments, 1.75-25.5 μm in diam. Skeletal hyphae brownish yellow, thick-walled with narrow but distinct lumen, unbranched, slightly twisted and sparsely branched towards extremities, otherwise cylindrical and straight, 3.75-5 μm in diam. Setae none. Basidia not observed. Basidiospores hyaline to slightly yellowish, almost rust brown in KOH, globose to sub-globose, thick-walled, smooth, nonamyloid, 3-4.5 × 3-5 μm.

DECAY : Causing a white pocket rot of heartwood. Sporophores appear up to a height of 2 m on the trunk.

SPECIMEN EXAMINED : On living *Dillenia pentagyna* Roxb. (DILLENIACEAE), near Perumpara (W), Sholayar forests, Malakkapara, Trichur Dist., Herb. CALI No. G343, 30-8-1986; on *Aglaia elaeagnoidea* (Juss.) Benth. (MELIACEAE). Tunacadavu, Parambiculam, Palghat Dist., Herb. FRI No. 7773, K. Bagchee, May 1947.

GEOGRAPHICAL DISTRIBUTION : North America, Nicaragua, Mexico, British Honduras, Panama, India.

REMARKS : Bakshi's (1971) description of *P. pseudosenex* based on a collection from Kerala by Bagchee, shows pores smaller than those of the present collection, although, both the collections were made from sites close to each other. Lowe (1952) reported the presence of hymenial setae in some of his collections of *P. pseudosensex* and he considered *Pyropolyporus troyanus* Murr. as a synonym. Later he (Lowe, 1957) described *P. pseudosenex* devoid of setae.

PHELLINUS PUNCTATUS (Fr.) Pilat

Atlas Champ. Europe 3 : 350, 1942.

Fig. 12. g-j; Pl., fig.

Polyporus punctatus Fr. Hymen. Europ. P. 572, 1874.

Poria punctata (Fr.) Karst., Bidr. Finl. Nat. Folk 37 : 83, 1882.

Fomitoporia obliquiformis Murr., North Am. Fl. 9 : 9, 1907.

Poria friesiana Bres., Ann. Mycol. 6 : 40, 1908.

P. sub-laevigata Clel. & Rodw., Pap. Roy. Soc. Tasm. P. 39, 1928.

Fuscoporia punctata (Fr.) G.H. Cunn., N.Z. Dep. Sci. Indstr. Res. Pl. Div. Bull. 73 : 11, 1948.

Fruitbody perennial, resupinate, inseparable from wood, thickened at the central region becoming plano-convex to pulvinate, up to 6 mm thick at the centre and receding towards the margin, woody hard. Pore surface amber yellow (4B6) to golden (4C6), colour fading towards the margin, smooth, cracking irregularly with age; margin pale yellow (3A3), adpressed and flat, finely velutinate; pores small and not visible to naked eye, round, regular, 5-6 per mm; pore mouth 90-120(180) µm wide; dissepiments 75-100 µm thick; pore tubes concolorous with the pore surface, stratified, each layer about 2mm long, old tubes filed with context tissue, tubes oblique. Context olive yellow (3C8), a thin layer present in between pore layers, up to 1 mm at the central portion, almost absent towards margin; of compact parallel hyphae.

Hyphal system dimitic. Generative hyphae rather frequent, hyaline, thin-walled, simple-septate, branched, easily deteriorating, 1.5-2µm in diam. Skeletal hyphae yellowish, thick-walled with broad to obliterate lumen, rarely branched, cylindrical or slightly twisted, ends hyaline and attenuated, 2-3 µm in diam. Setae none. Cystidia not observed. Basidia short, broadly clavate, 4-spored, 10-12 × 7-8 µm; sterigmata 2-3 µm long, straight. Basidiospores globose to slightly sub-globose, hyaline while young, slightly yellowish when mature, smooth, thick-walled, dextrinoid, 5-6 µm in diam. In case of globose spores, otherwise 5-6 × 5-5.5 µm.

DECAY : White rot, sapwood becoming brittle, can be easily crushed into a white powdery mass, heartwood intact.

SPECIMEN EXAMINED : On a dead branch of *Santalum album* Linn. (SANTALACEAE), Botanical Garden, Calicut University Campus, Malappuram Dist., Herb. CALI No. G61, 20-6-1984.

GEOGRAHPICAL DISTRIBUTION : East Africa, French Antiles, French Guiana, Portugal, Finland, Bulgaria, France, USSR, USA, India.

REMARKS : The present collection, when compared with descriptions of *P. punctatus* from temperate countries (Baxter, 1926; Lowe, 1958, 1963, 1966; Gilbertson, 1979; Bondarstev, 1953; Ryvarden, 1978; Ireneia, 1983) showed slightly wider pores, broader skeletal hyphae and smaller spores. It is more akin to the sporophores reported from French Antilles and Guiana (David & Rajchenberg, 1985). According to David et al. (1982) the presence of thick-walled setae like structures reported in *P. punctatus* by Bondarstev (1953) is not characteristic of *P. punctatus*, but reminds one of a setuloid species *P. psedopunctatus.*

This is the first report of *P. punctatus* from India and only a single collection could be made during the present study.

PHELLINUS RIMOSUS (Berk.) Pilat
Ann. Mycol. 38 : 80, 1940.

Fig. 12. k-n; Pl. fig.

Polyporus rimosus Berk., Lond. J. Bot. 4 : 54, 1854.
P. igniarius var. *Scaber* Berk., Ann. Nat. Hist. 3 : 324, 1839.
Fores rimosus (Berk.) Cke., Grevillea 14 : 18, 1885.
Polyporus pappianus Bres., A. Roy. Inst. Bot. Roma. 6 : 178, 1896.
Xanthochrous rimosus (Berk.) Pat., Essai Tax. P. 101, 1900.
Fomes versicolor Henn., Hedwigia 42 : 75, 1903.
F. dialeri Bres. & Torr., Broteria 4 : 218, 1905.
Pyropolyporus cedrelae Murr., North Am. Fl. 9 : 105, 1908.
Fomes aulaxinus Bres., Ann. Mycol. 10 : 497, 1912.
F. scaber (Berk.) Lloyd, Syn. Genus *Fomes* p. 249, 1915.
F. tepparii Lloyd, Ibid. P. 256, 1915.

Fruitbody perennial, epixylous, solitary, rarely imbricate and confluent, pileate and sessile, attached with a broad lateral base, conchate to dimidiate, applanate to ungulate or triquetrous, hard woody, heavy, 8-17. × 5-9.5 × 2.5-9 cm. Pileus surface brownish grey. (8E2) to bluish grey (22F2) or dark grey (1F1) to black, birch grey (5C2) near margin; glabrous, rough, concentrically sulcate, heavily crusty, often breaking into irregular blocks; margin slightly tomentose soon becoming glabrescent, thick and rounded. Pore surface yellowish brown (5E8) to olive brown (4E7) or golden brown (5E7) to coffee (5F7), slightly rough, shiny while fresh; pores small but visible to naked eye, extending to the margin, round, regular, 4-6 per mm; pore mouth 90-150 µm wide; dissepiments 60-150(210) µm thick; pore tubes stratified, slightly darker than the pore surface, 2-4 mm long in each stratum, old tubes filled by context-tissue. Context chinese yellow (4B7) to yellowish brown (5D8), becoming hard and black near the crust, fibrous, 1-2.5 cm thick, of compact parallel hyphae.

Hyphal system dimitic. Generative hyphae hyaline, thin-walled, sparsely simple-septate, branched, easily breaking, more in dissepiments, 1.5-3(3.75) µm in diam. Skeletal hyphae yellowish brown, thick-walled with broad lumen, simple-septate, unbranched or rarely branched at extremities, common, 2.5-6.25 μm in diam. Setae none. Basidia slightly pyriform to sub-globose or broadly clavate, 2 or 4-spored, 10.5-13.5 × 4.5-6.75 μm; sterigmata 1.5-3 μm long. Basidiospores brownish yellow, reddish in KOH, thick-walled, ovoid to sub-globose to globose, smooth, 4-6, × 4-5.5 μm; spore print apricot (5B6).

DECAY : White pocket rot of heartwood, sapwood usually not attacked.

SPECIMEN EXAMINED : On living trunks of *Artocarpus heterophyllus* Lam. (MORACEAE), Divisional Forest Office campus, Nilambur, Malappuram Dist., Herb. CALI No. G152, 7-3-1985; G153, 7-3-1985; Calicut University Campus, Malappuram Dist., Herb. CALI No. G188, 14-6-1985; Ayyanthole, Trichur Dist., Herb. CALI No. G280, 30-3-1986; on *Ficus benghalensis* Linn. (MORACEAE), Chakkumpuzha, Palai, Kottayam Dist., Herb. CALI No. G124, 3-7-1983; on living turnks of *Terminalia cremulata* Roth. (COMBRETACEAE), Panangode Block, Nilambur Range, Malappuram Dist., Herb. FRI No. 4286, K. Bagchee, 19-1-1946.

GEOGRAPHICAL DISTRIBUTION : Australia, New Zealand, Tasmania, Phillippines, Sri Lanka, East Africa, Mauritius, Portugal, Turkey, North America.

REMARKS : *Phellinus rimosus* was earlier reported from India as *Fomes rimosus* by Bagchee et al. (1954), while Bakshi (1971) described it as *F. Robineae* (Murr.) Sacc, & D. Sacc. According to Gilbertson (1979) *Phellinus robineae* (Murr.) A Ames (=*F. Robineae*) is host specific (*Robinia* sp.) and has smaller pores and generally sporophores more applanate than those of *P. rimosus.*The present collection shows close similarly to both the earlier reports from India, and the East African collections of Ryvarden & Johansen (1980). The New Zealandean sporophores (Cunningham, 1965) differs in having wider pore mouth and more elliptical spores.

The Taxonomy of the three closely related polypores, viz. *Phellinus badius* (Berk.) G.H. Gunn., *P. rimosus* and *P. robineae* is in a confused state as these names have been variously interchanged in the past (Gilbertson, 1979). According to Gilbertson (1979) *P. robineae* is host specific, while *P. badius* is represented only by the type specimen and *P. rimosus* is a non-host specific, cosmopolitan species.

In Kerala, *P. rimosus* is quite common on living Moraceae members. It causes white pocket rot initially but later the heartwood is transformed into a white spongy mass. Just below the basidiocap, a large open wound reveals the decaying heartwood. The basidiocarp is locally used in treating mumps. It is made into a paste and smeared on the swollen parts around the jaws.

PHELLINUS SENEX (Nees & Mont.) Imaz.

Bull Govt. For. Exp. Sta. Tokyo 57 : 115, 1952. Fig. 13. a-c; Pl.

Polyporous senex Nees & Mount., Ann. Sci. Nat. 2,5 : 70, 1836.

P. korthalsii Lev., Ibid. 3, 2 : 190, 1844.

P. auriculaeformis Lev., Ibid, 3, 2 : 194, 1844.

Fomes senes (Nees & Mount.) Cke., Grevillea 13 : 118, 1885.

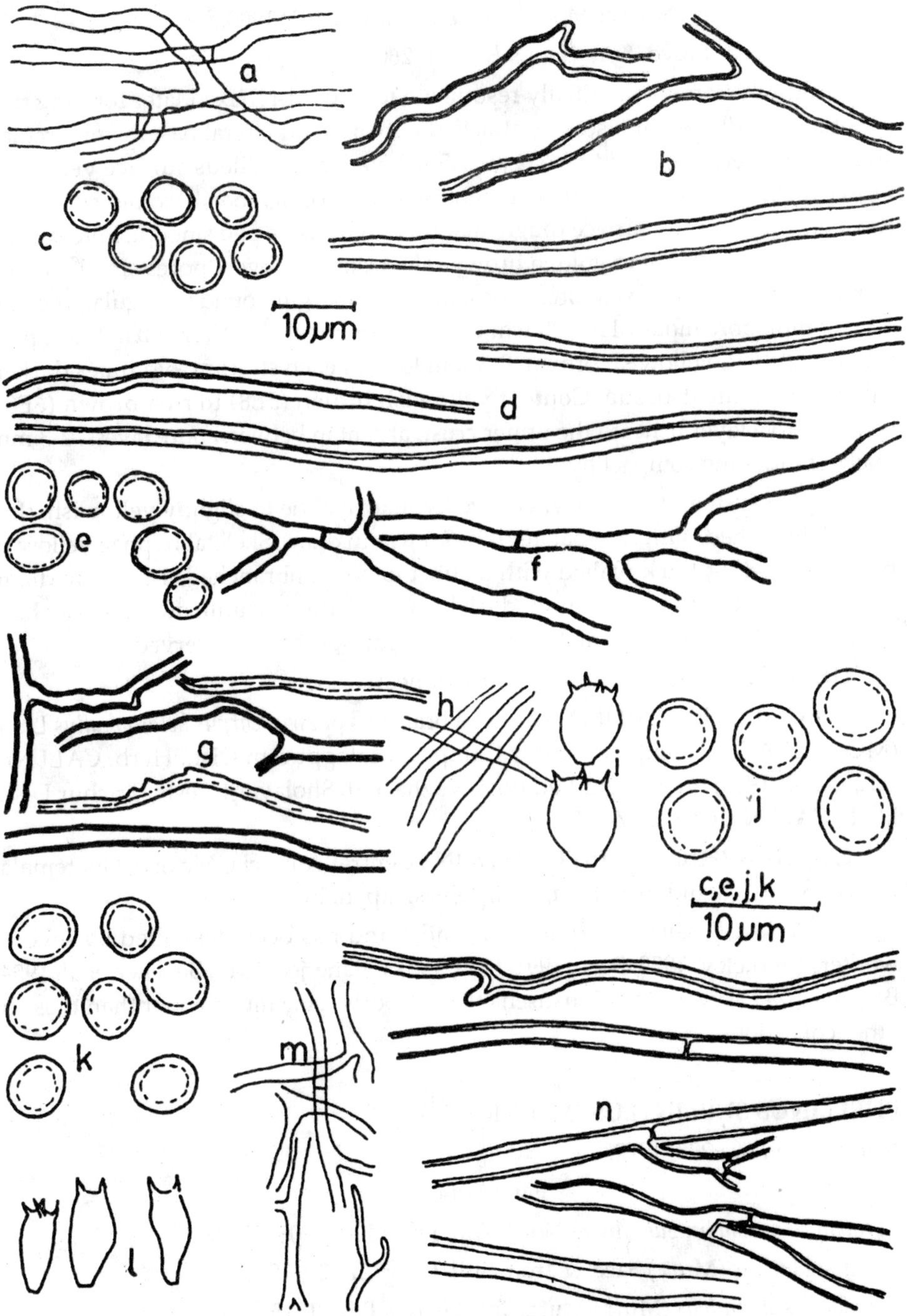

Fig. 12 : a-c *Phellinus nilgheriensis* : a-generative hyphae; b-skeletal hyphae; c-basidiospores. d-f *Phellinus pseudosenex* : d-skeletal hyphae; e-basidiospores; f-generative hyphae. g-j *Phellinus punctatus* : g-skeletal hyphae; h-generative hyphae; i-basidia; j-basidiospores. k-n *Phellinus rimosus* : k-basidiospores; l-basidia; m-generative hyphae; n-skeletal hyphae.

Xanthochrous senex (Nees & Mont.) Pat., Essai Tax. P. 101, 1900.

Fomes hawaiensis Lloyd, Syn. Genus *Fomes* P. 260, 1915.

Fruitbody perennial, initially resupinate to effused-reflexed later forming true pileate fruibody, solitary, sessile, attached with a broad lateral base, semi-circular, slightly triquetrous, woody, heavy; 11.5 × 7.5 × 2.5 cm. Pileus surface yellowish brown (5F6) to dark brown (6F6), tomentose while young, soon becoming glabrous, concentrically sulcate, surface rough; margin tough, entire and smooth. Pore surface yellowish brown (5E5) to golden brown (5D7), smooth, even; pores small and not visible to unaided eye, extending almost up to the margin, round to angular, regular, 6-7 per mm; pore mouth 110-130 µm wide; dissepiments 50-90(120) µm thick; pore tubes stratified, yellowish brown, 2-3 mm long in each stratum, old tubes darker, filled with context-tissue. Context yellowish brown (5E8) to rust brown (6E8), confined to a layer beneath the upper crust, absent in between pore layers, 2-3 mm thick, of radiating compact hyphae.

Hyphal system dimitic. Generative hyphae hyaline to slightly yellowish, thin walled, branched, simple-sepatate, 1.75-2.5 µm in diam. Skeletal hyphae yellow to brownish yellow, thick-walled with distinct lumen, unbranched, 4-5 µm in diam. Setae frequent in the hymenium; dark brown, acute to acuminate, thick-walled, straight, 15-25(30) × 5-75 µm. Basidia and basidiospores not observed.

DECAY : Causing white stringy rot of heartwood.

SPECIMEN EXAMINED : On dead branches of *Artrocarpus heterophyllus* Lam. (MORACEAE), Calicut University Campus, Malappuram Dist., Herb. CALI No. G126, 20-9-1984; on unidentified host, Vazhachal, Sholayar forests, Trichur Dist., Herb. CALI No. G83, 26-7-1984.

GEOGRAPHICAL DISTRIBUTION : Central Amercia, Mexico, Guatemala, Africa, New Zealand, Australia, Philippines, Japan, India.

Remarks : *P. senex* is common in India and has been described by several workers (Berkeley, 1939; Bose, 1991-1928, 1937a; Banerjee, 1947; Bagchee et al. 1954; Bakshi, 1971). Ryvarden & Johansen (1980) reports setae much wider than those of other collections.

PHELLINUS SUBLINTEUS (Murr.) Ryv.

Norw. J. Bot. 19 : 235, 1972.

Fig. 13. d-e; Pl. Fig.

Pyropolyporus sublinteus Murr., North Am. Fl. 9 : 110, 1908.

Fomes sublinteus (Murr.) Sacc. & Trott. Sull. Fung. 21 : 29, 1912.

Fulviformes sublinteus (Murr.) Murr., Trop. Polyp. P. 86, 1915.

Fruitbody annual to perennial, solitray to imbricate, pileate, sessile, attached laterally with a broad base; dimidiate to semi-circular, applanate, hard corky while fresh as well as dry; light in weight; 12-18 × 11-13 × 0. 6-1 cm. pileus surface drak grey (1F1) to dark brown (6F6); uneven and rough, glabrous, concentrically zonate, crusty towards base, often wrinkled; margin smooth, rounded, entire. Pore surface yellow to yellowish brown (5F8) to raw umber (5F8), rough, growth not uniform towards the base; pores extending almost up to the marigin, small but visible to naked eye, angular, 5-6 sided, regular, 4-6 per mm; pore mouth 130-180 μm wide ; dissepiments 40-70 μm thick; pore tubes raw umber (5F8), 5-7 mm long, sunk into equal depths, shallow to wards the margin. Context dark yellow (4C8), fibrous, homogeneous, of compact parallel hyphae.

Hyphal system dimitic. Skeletal hyphae yellow, slightly thick-walled with broad lumen, less frequently short branched, simple-septate, cylindrical and straight, 5-6.25 μm in diam. Generative hyphae not observed. Setae none. Basidia not observed. Basidiospores yellowish brown, dark brown in KOH, globose to sub-globose, thick-walled, smooth, 5-6 × 4-5. 5 μm.

DECAY : White rot, the decaying wood first becoming fibrous and later an amorphous mass.

SPECIMEN EXAMINED : On decomposing forest logs, Divisional Forest office campus, Nilampur, Malapuram Dist., Herb. CALI No. G279, 7-3-1085

GEOGRAPHICAL DISTRIBUTION : North America, Mexico, Jamaica, Cuba, India, East Africa.

REMARKS : The present collection matches well the with the earlier Indian collections (Bakshi, 1971) as well as the East African collections of Ryvardan & Johansen (1980) in all macro- and microcharacters. However, Murrill (1915) reported globose Spores and Lowe (1957) reported ovate spores for Mexican and North American collections respectively.

PHELLINUS TROYANUS (Murr.) Ganesh & Leelavathy comb nov.

Bas. Pyropolyporus troyanus Murr., Mycologia 2 : 196, 1910.

fig. 13. f-h ; Pl. , fig

Fomes tryoyanus (Murr.) Sacc. & Trott., Syll. Fung. 21 : 289, 1912

Fulvifomes troyanus (Murr.) Murr., Trop. Polyp. P. 87. 1915.

Fruitbody perennial, solitary, pileate, sessile, attached with a broad lateral base, semi-circular to dimidiate, rarely irregular, ungulate or triquetrous, woody and heavy; 7-21 × 4–10.5 × 3.5–7 cm. Pileus surface nutria brown (5F3) to sepia brown (5F4), greyish brown (6E3), to black, dark brown (6F8) near margin ; rough and uneven, glabrous, concentrically rimose, crusty, radially cracking ; margin rounded, smooth. Pore surface dark yellow (4C8) to brownish yellow (5C7) or light brown (5D8), yellowish brown (5D8) near margin ; pores minute and not visible to naked eye, absent near margin, slightly angular to round, 8-10(12) per mm; pore mouth 60-85 μm wide; dissepiments 35-50 μm thick; pore tubes golden yellow (5B7) to oxide yellow (5C7) towards upper crust, stratified, 2-3 mm long in each stratum. Compact, concolorous with the pore tubes, present in between pore layers, 1-2 mm thick, of compact radiating hyphae, thin, black lines presest in the context between pore layers.

Hyphal system dimitic. Generative hyphae hyaline, slightly thick-walled, branched, simple-septate, infrequent, 2. 5-3 μm in diam. Skeletal hyphae yellow to brownish yellow, thick-walled, usually unbranched, sometimes weakly branched near extremities, main stem cylindrical and equal in width, 2. 5-5 μm in diam. Satae none. Basidia not observed. Basidiopores yellow to yellowish brown, dark brown in KOH, sub-globose, thick-walled, smooth, (3)4-5 × (3) 3.5-4 μm.

DECAY : White rot of heartwood.

SPECIMEN EXAMINED : On dead logs of *Dimocarpus longan* Lour. (SAPINDACEAE), Vengoli peak, Parambiculum forests, Palghat Dist., Herb. CALI No. G204, 11-7-1985; on unidentified forest logs, Malakkapara, Sholayar forest, Trichur Dist., Herb. CALI No. G183, 10-6-1985.

GEOGRAPHICAL DISTRIBUTION : Jamaica, India.

REMARKS : This is the first report on the occurrence of *P. troyanus* outside of Jamaica. The present collection shows close similarity with Lowe's redescription (Lowe, 1957) of Murrill's type collections except for the smaller size of pores in the present material.

This species has a dimitic hyphal system with coloured skeletal hyphae darkening in KOH and clampless generative hyphae which are features typical of the genus *Phellinus*. The transfer of the species to *Phellinus* is therefore proposed here.

PHELLINUS WAHLBERGII (Fr.) Reid.

Contr. Bolus Herb. 7 : 97, 1975.

Fig. 13. i-k; Pl. Fig.

Trametes wahlbergii Fr., Kung. Vet. Akad. Handb. P. 131. 1948.

Polyporus zealandicus Cke., Grevillea 8 : 55, 1897, non Cke. 1888.

P. victoria Berk. ex Cke., Ibid. 10 : 103, 1882.

Fomes zealandicus Cke., Ibid, 14 : 18, 1885.

Poria victoriae (Berk.) Cke., Ibid. 14 : 111, 1886.

Pyropolyporus robinsoniae Murr., North Am. Fl. 9 : 108, 1908.

Poria uncinata Weir ex Clel. & Rodw, Pap. Roy. Soc. Tasm. P. 39, 1928.

Fomes senex var. *Hamatus* Corner, Trans. Brit. Mycol. Soc. 17 : 75, 1932.

Fomes hamatus (Corner) Imaz., J. Jap. Bot. 16 : 586, 1940

Fuscoporia uncinata (Weir.) G.H. Cunn., N.Z. Dep. Sci. industr. Res. Pl. Dis. Div. Bull. 73 : 3, 1948.

Phellinus zealandicus (Cke.) G.H. Cunn., Ibid. 164 : 223, 1965.

Fruitbody perennial, initially slightly effused-reflexed, soon becoming pileate, sessile, attached with a broad lateral base, semi-circular, applanate to broadly ungulate, woody hard but light in weight; 26.5 × 12 × 6 cm or more. Pileus surface rust

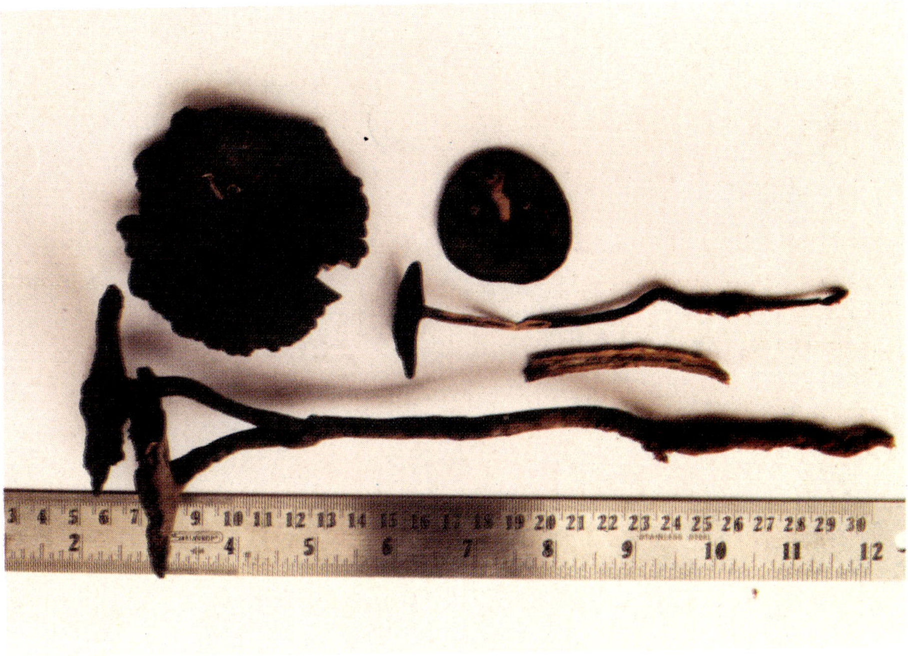

1. Amuroderma rugosum—habit

2. Amuroderma rugosum—pileus surface

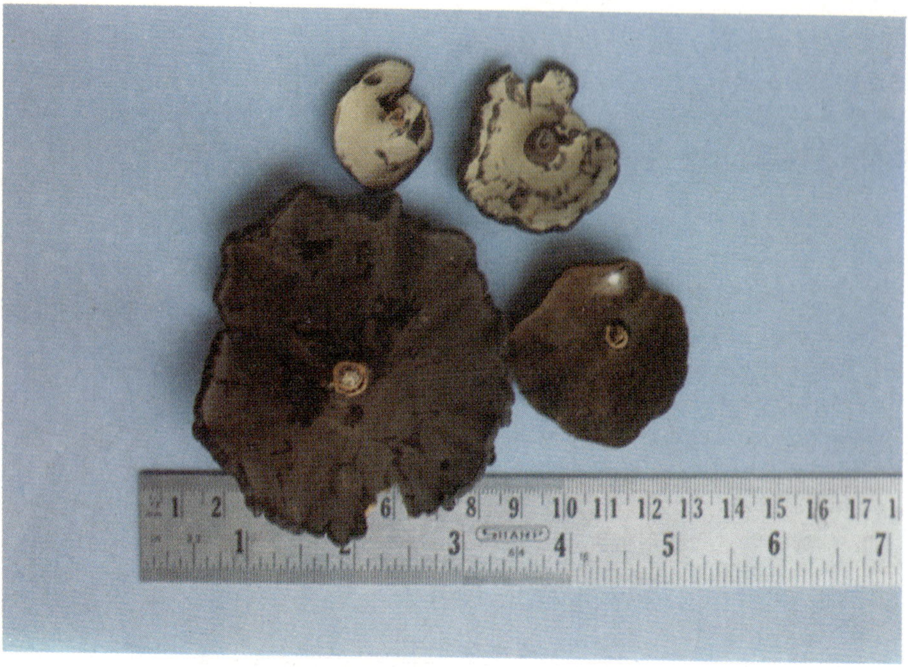

3. Amuroderma rugosum—pore surface

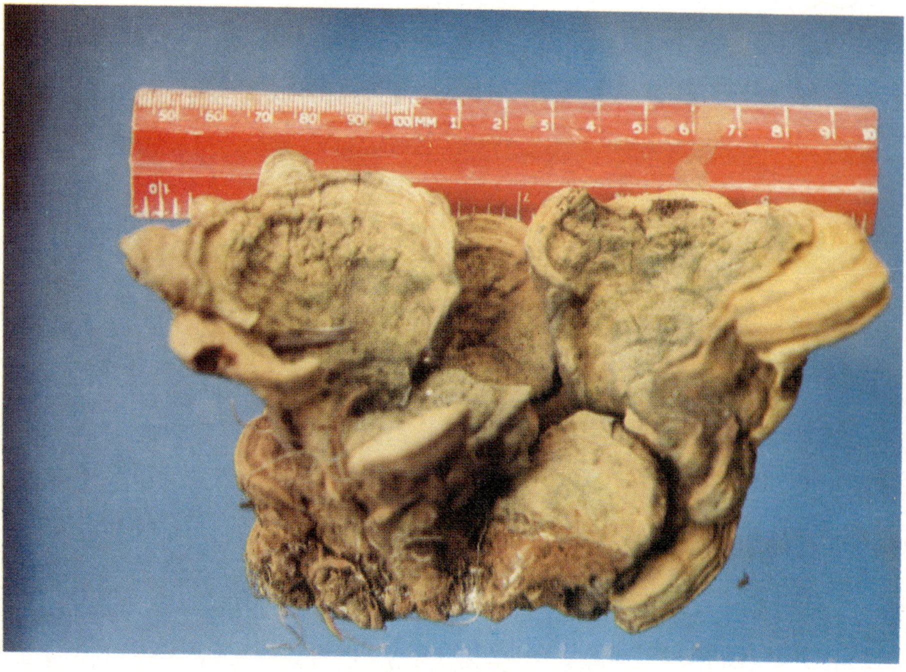

4. Ganoderma australe—habit

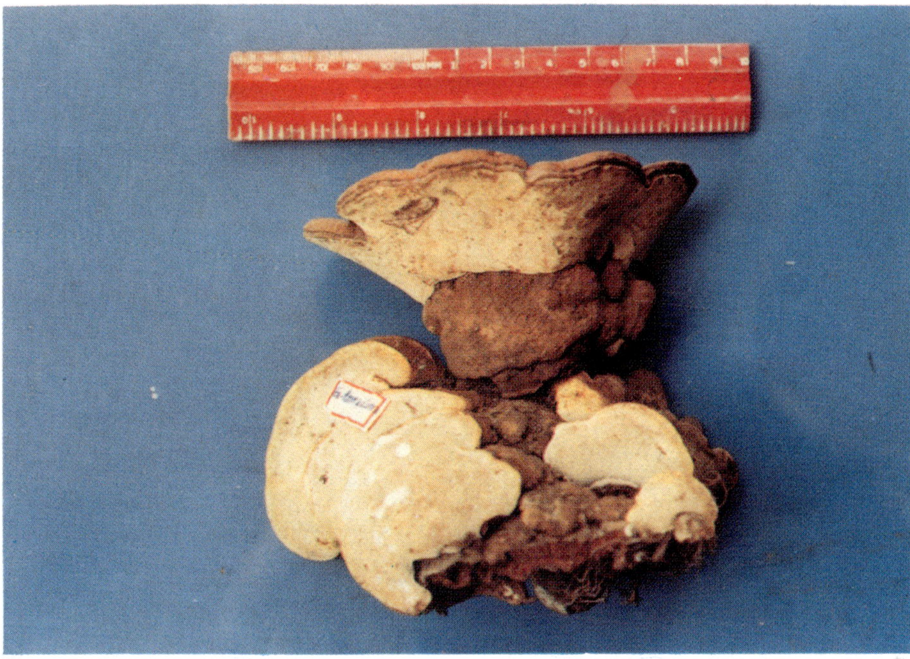

5. Ganoderma austrrale—pore surface

6. Ganoderma lucidum—fruit bodies at the base of host tree.

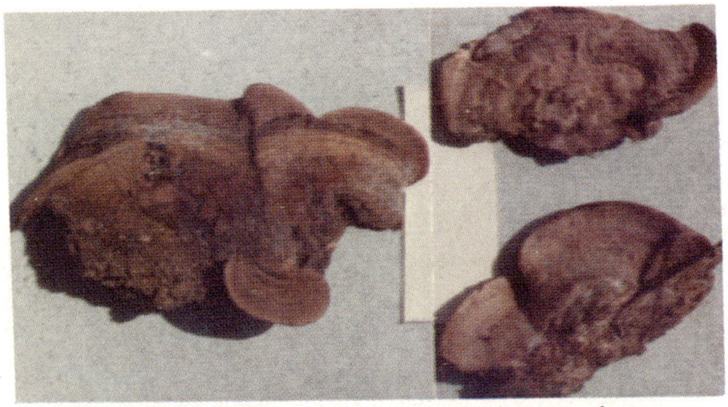

27.(a) Microporus affinis—pileus and pore surface

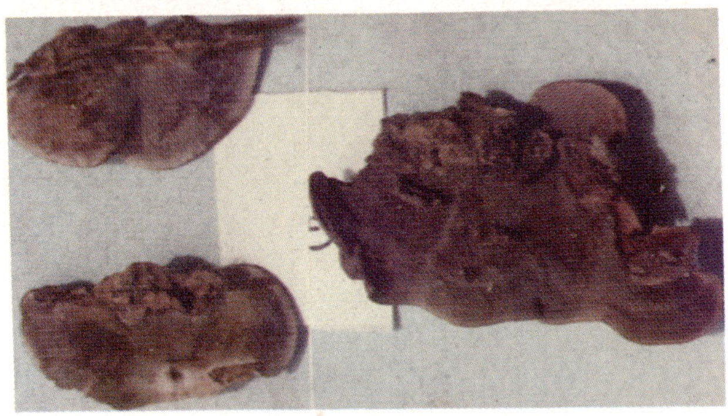

27.(b) Nigroporus vinosus—pileus surface

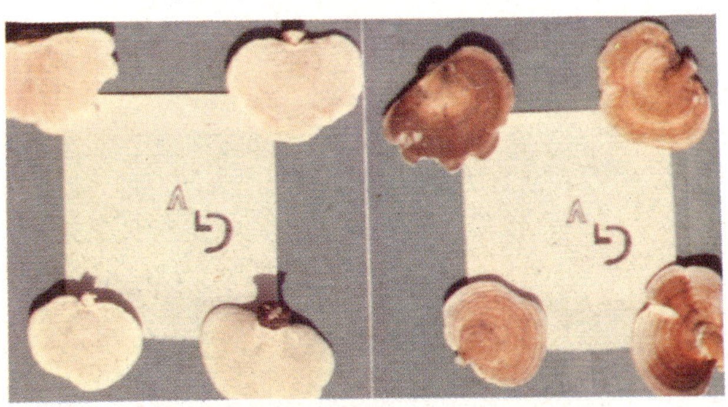

27.(c) Nigroporus vinosus—pore surface

28. Perenniporia ochroleuca—pileus and pore surface

29. Pycnoporus sanguineus—habit

30. Polyporus arcularius—habit

31. Polyporus sp.—pileus and pore surface

32. Rigidoporus lineatus—habit

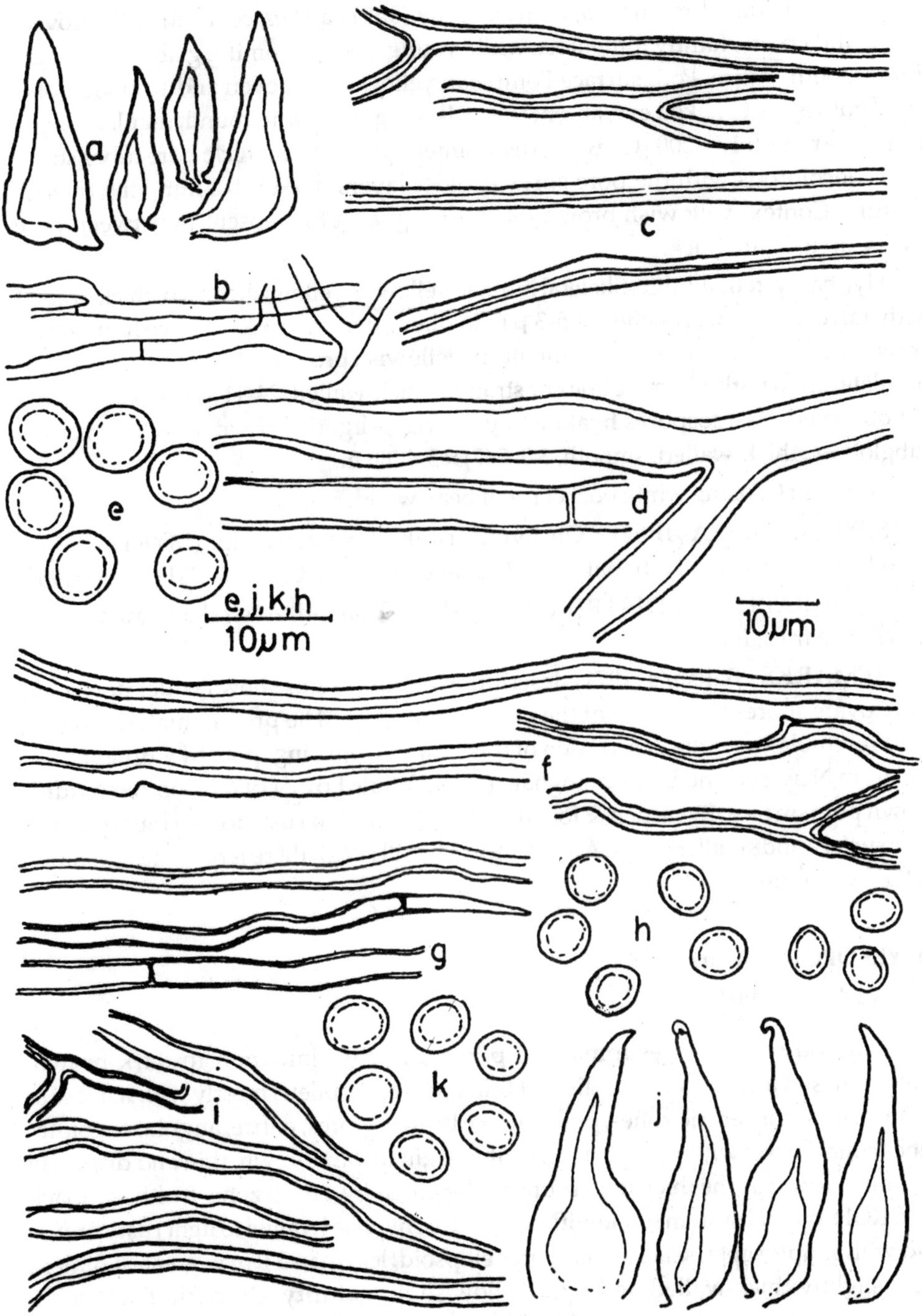

Fig. 13 : a-c *Phellinus senex* : a-setae; b-generative hyphae; c-skeletal hyphae.
d-e *Phellinus sublinteus* : d-skeletal hyphae; e-basidiospores.
f-h *Phellinus troyanus* : f-skeletal hyphae; g-generative hyphae; h-basidiospores.
i-k *Phellinus wahlbergii* : i-skeletal hyphae; j-setae; k-basidiospores.

brown (6E8) to dark brown (6F8) with alabaster (5B2) patches; concentrically groved, rough and crusty, faintly tomentose while young, soon becoming glabrous; margin thin, smooth, entire. Pore surface Pompeian yellow (5D6), even and smooth; pores small but visible to naked eye, absent towards margin, ovate to round, regular, 4-5 per mm; pore mouth 170-200 μm wide; dissepiments 60-100 μm thick; pore tubes stratified, concolorous with the pore surface, older layers darker, 1-2 mm long in each stratum. Context yellowish brown (5D8) to cognac (6E7), absent in between pore layers, up to 5 mm thick.

Hyphal system dimitic. Skeletal hyphae yellow to yellowish brown, thick-walled with narrow lumen, branched, 1.5-3 μm in diam., common. Generative hyphae not observed. Setae present in the hymenium; yellowish brown to brown, ventricose to subulate, often with a hamate apex or straight, thick-walled, 24-41 × 6-10 μm. Basidia not observed. Basidospores hyaline to yellowish, slightly brown in KOH, ovoid to subglobose, thick-walled, smooth, 3.5-5 (7) × 3-4.5 μm.

DECAY : Causing white pocket rot of heartwood.

SPECIMEN EXAMPED : On living trunk of *Syzygium* sp. (MYRTACEAE), Valakkad, Silent Valley National Park, Palghat Dist., Herb. CALI No. G112, 30-6-1984.

GEOGRAPHICAL DISTRIBUTION : New Zealand, Australia, New Guinea, America, Africa, India.

REMARKS : Bakshi's (1971) description of this fungus shows wider pores and more ovate spores than those of the present collection. The present material shows close resemblance to the description of this fungus from Singapore by Corner (1932) and from New Zealand by Cunningham (1965). While Lowe (1957) reported reddish brown pore surface, Ryvarden & Johansen (1980) reported rusty to chestnut coloured pore surface and smaller pores. According to Murrill (1915) there is no setae ('cystidia') in the hymenium.

PHYLLOPORIA Murr.

Torreya 4 : 141, 1904.

Fruitbody annual, resupinate to pileate, pileus cinnamon to dark brown, tomentum soft and thick over a distinct thin dark, zone, pileus mostly with narrow to wide, sulcate concentric zones, pore surface brown, pores, entire, angular to round, tubes concolorous with pore surface, context light to dark brown, thin and distinctly delimited towards the thick tomentum on the pileus by a dark zone, fruitbody black with KOH, hyphal system monomitic, generative hyphae hyaline to light rusty brown and with simple septa, setae none, spores ellipsoid, less than 5 μm in greatest dimension, slightly thick-walled and light yellowish in maturity. On deciduous wood, often on living bushes or remarkably thin dead branches on living trees, or on living leaves. Mainly tropical genus.

Types species : *Phylloporia parasitica* Murr.

PHYLLOPORIA CHRYSITA (Berk.) Ryv.

Norw. J. Bot. 19 : 235, 1972 Fig. 32 a-b; Pl. Fig.

Polyporus chrysites Berk., Hooker J. Bot. 8 : 233, 1856 (K)

Inonotus corrasus Murr., Bull., Torr. Bot. Cl. 31 : 598, 1904.

Fruitbody annual, pileate and sessile, densely imbricate on the substrate; pileus fan shaped 1.0-1.5 × 1.5-1.8 cm, up to 4mm thick at the base, margin thin, surface golden yellow (5B7) to light brown (6D8) to rust brown (6E8), margin paler, light brown (7D8), zonate, velvety hairy below which there is a dark line; pore surface light brown, concolorous to pileus surface, margin sterile, pores circular to oval, small, 6-7 per mm; Pore tubes up to lmm deep, brown as pore surface. Context 2-3 mm thick, dark brown (7F8) separated from the tomentum by a dark line.

Hyphal system monometic with simple septate generative hyphae, rusty brown, 3-5 μm broad, slightly thick-walled. Basidia clavate, with 4 strigmata. Basidiospores subglobose, 3-3.5 × 2-2.25 μm, light yellow, slightly thick-walled, smooth. Setae and setal hyphae absent.

SPECIMEN EXAMINED : On the surface of unidentified standing deciduous tree (dead?) busting out through bark from base up to 20 feet in a wooded area, Calicut University Campus, Malappuram Dist. Kerala, India, 1-10-1989, Herb. CALI. No. F549.

GEOGRAPHICAL DISTRIBUTION : Florida, Cuba, Jamaica, West Indies, Venezuela, Brazil, Indonesia and Philippines.

REMARKS : This seems to be the first and only report of the genus *Phylloporia* from India, although it is widely distributed in the tropics. The densely imbricate habit covering large areas of the stem and exposed roots of a standing tree and the rusty brown, velvety surface of the basidiomes make it quite conspicuous.

POLYPORACEAE CORDA

Icon. Fung. 3 : 49, 1839.

Fruitbody annual to perennial, stipitate, pileate or resupinate, colour variable from white, cream, ochre, brown, red, violet or bay to black. Hymenophore mostly poroid, but also lamellate, daedaleoid or labyrinthine, in variable colours, shallow to deep and stratified. Context in variable colours, homogeneous to duplex. Hyphal system mono-, di- or trimitic, the latter mostly in long-lived or perennial species; generative hyphae with simple-septa or more commonly with clamps; skeletal and binding hyphae hyaline or coloured and in the latter case fruitbodies also coloured at least in section. Cystidia absent or present. Basidiospores of variable shape, usually smooth, more rarely ornamented, with a simple and single wall, hyaline to coloured, thin to-thick-walled, usually negative in Melzer's but also amyloid or dextrinoid. Mostly on dead wood, more rarely on living plants or on the ground. Cosmopolitan family with numerous genera and species.

Type genus : *Polyporus* Fr.

26 genera and 53 species have been recorded from Kerala.

KEY TO GENERA

1. Fruitbody centrally or laterally Stipitate.............................. 2
1′. Fruitbody resupinate, effused-reflexed or pileate but not stipitate.............. 7
2. Context homogeneous; pores round to angular.............................. 3
2′. Context duplex; pores daedaleoid to irpicoid.............................. *Abortiporus*
3. Pores small, round to angular, scattered; pore mouth entire.............................. 4
3′. Pores large, hexagonal in radiating rows; pore mouth fimbriate ...... *Favolus*
4. Spores amyloid.............................. *Amylosporus*
4′. Spores nonamyloid.............................. 5
5. Fruitbody very thin; hyphal system trimitic.............................. 6
5′. Fruitbody fairly thick; hyphal system dimitic.............................. *Polyporus*
6. Spores elliptical; coralloid elements present in dissepiments....... *Microporus*
6′. Spores globose to sub-globose; coralloid elements nil.............. *Microporellus*
7. Cystidia encrusted or encrused skeletal hyphae projecting into the hymenium 8
7′. Cystidia usually none; if present never encrusted.............................. 17
8. Hyphal system monomitic, hyphae simple-septate.............................. 9
8′. Hyphal system di-or trimitic.............................. 10
9. Pore surface white to cream or pale brown; rather soft and fragile, if hard, then with very small pores and pileate; cystidia only of one type.............................. *Oxyporus*
9′. Pore surface vivid orange, pinkish, brownish, ochraceous to grey or black; encrused cystidia and smooth, small, mammilate cystidia present.............................. *Rigidoporus*
10. Spores minute, allantiod.............................. 11
10′. Spore cylindric to ellipsoid to globose.............................. 12
11. Fruitbody resupinate to narrowly reflexed; context white, very thin.............................. *Skeletocutis*
11′. Fruitbody effused-reflexed to semi-pendent; context orange, prominent and fairly thick.............................. *Piloporia*
12. Fruitbody strictly resupinate; context white, Extremely thin....... *Junghuhnia*
12′. Fruitbody effused-reflexted; context pale to coloured, thick.............................. 13
13. Hymenophore poroid.............................. 14
13′. Hymenophore lamellae to daedaleoid.............................. *Lenzites*
14. Pores hexagonal.............................. *Hexagonia*
14′. Pores not so.............................. 15
15. Fruitbody pileate, rarely effused-reflexed; abhymenial hairs forming a coarse tuft.............................. *Trichaptum*
15′. Fruitbody effused-reflexed; abhymenial hairs usually absent or when present smooth and velutinate.............................. 16
16. Pileus surface and context yellow, reddish in KOH.............................. *Flavadon*

16′. Pileus surface and context white to grey, negative in KOH.................... *Irpex.*
17. Hyphal system monomitic, hyphae nodose-septate.. 18
17′. Hyphal system di-or trimitic.. 19
18. Context white and homogeneous, in one species separated from tubes by a thin black zone.. *Bjerkandera*
18′. Context duplex.. *Tyromyces*
19. Spores truncate.. 20
19′. Spores smooth-walled, ellipsoid to ovate or globose.................................... 22
20. Fruitbody brown to reddish brown, pore surface grey to brown to reddish brown, skeletal hyphae coloured.. 21
20′. Fruitboday plae orange to orange white, pore surface creamy white to organge white, skeletal hyphae hyaline.. *Perenniporea*
21. Fruitbody reddish, pore surface reddish brown, skeletal hyphae orange to red in mass... *Pyrofomes*
21′. Fruitbody in shades of brown, pore surface grey to brown, skeletal hyphae yellow to pale brown.. *Lowepores*
22. Context brightly coloured.. 23
22′. Context white, cream or ochre 24
23. Pileus surface reddish; context cinnabar-red; hyphal system trimtic.. *Pycnoporus*
23′. Pileus surface vinaceous brown; context reddish brown; hyphal system dimitic.. *Nigroporus*
24. Fruitbody perennial, mostly with a distinct crust at least at the base; woody hary.. *Fomitopsis*
24′. Fruitb annual to biennial; pileus surface smooth to tomentose or hirsute, mostly coriaceous to corky... 25
25. Context brownish, darker in KOH.. *Coriolopsis*
25′. Context white to cream; negative in KOH or faintly yellowish, never dark brown.. *Trametes*

ABORTIPORUS Murr.

Bull. Torrey Bot. Club 31 : 421, 1904.

Fruitbody annual, sub-stipitate and infundibuliform or sessile, dimidiate. Pores angular to daedaleoid. Context white to pale buff, duplex. upper layer soft and spongy, lower layer firm and fibrous. Hyphal system monomitic; generative hyphae with clamps. Chlamydospores present in upper context. Cystidia present or absent. Basidiospores hyaline, smooth, sub-globose to ellipsoid, negative in Melzer's.

Type species : *Abortiporus distortus* (Schw.) Murr.

One species has been collected from Kerala.

ABORTIPORUS BIENNIS (Bull. ex Fr.) Sing.
Mycologia 36 : 68, 1944., Fig. 14 a Pl. Fig.
Boletus biennis Bull, Herb. Fr. Pl. 449, Fig., 1. 1790.
Daedalea biennis (Bull.) ex Fr. Syst. Mycol. 1 : 332, 1821.
Sistotrema rufascens Pers., Syn. Fung. P. 550, 1801.
Polyporous rufascens (Pers.) Fr., Syst. Mycol. 1 : 351, 1821.
Boletus distortus Schw., Syn. Fung. Carol. Super 2 : 71, 1822.
Polyporus laciniatus Pers., Mycol. Eur. 2 : 48, 1825.
Boletus distortus Schw. ex Fr., Elench. Fung. 1 : 79, 1828.
Polyporus distortus (Schw. ex Fr.) Fr., Ibid. 79. 1918.
P. biennis (Bull. ex Fr.) Fr., Epicrisis p. 433, 1838.
P. acanthoides Fr., Ibid. P. 438, 1838.
Daedalea hymenopus Brig., Atti. Acad. Sci. Nat. Napoli 7 : 97, 1847.
Polyporous proteiporus Cke., Grevillea 12 : 15, 1883.
P. sericellus Sacc., Syll. Fung. 7 : 77, 188, non Lew., 1846.
P. novaeguineensis Henn., Fl. Kaisel with. Land 6, 1889.
Daedalea incarnato-albida Chobat & Martin, Beih, Bot. Centralbl. P. 100, 1891.
D. delicatissimia Speg., Ann. Mus. Nat. Hist. B. Aires 6 : 173, 1898.
D. boroniensis Speg., Ibid. 3, 1 : 52, 1902.
Abortiporus distortus (Schw. ex Fr.) Murr., Bull. Torrey Bot. Club 31 : 421, 1904.
Irpicium ulmicola Bref., Unter Gesamtgeb Mykol. 15 : 143, 1912.
Heteroporus biennis Pilat, Beih. Bot. Centralbl. 52 : 63, 1934.
Polyporus balloni Lloyd, Mycol. Notes 69 : 1191, 1923.
Phaeolus biennis Pilat, Beih. Bot. Centralbl. 52 : 63, 1934.
Polyporus gratzianus Murr., Florida Acad. Sci. J. 8 : 197, 1945.

Fruitbody annual, effused-reflexed to pileate, laterally sub-stipitate to sessile, attached with a narrow converging base, rarely with a broad base; semi-circular to reniform or flabelliform, applanate, conchate, soft spongy while fresh, rigid when dry; 2.5-3.2 × 1.8-3 × 0.4-0.7 cm. Pileus surface yellowish brown (5E4) to dark blonde (5D4), margin white with an orange tinge, light yellow (4A4) when dry; matted tomentose, even, azonate, slightly wrinkled when dry; margin smooth and entire Pore surface white with a greyish orange (5B4) tinge while fresh, cream (4A3) when dry; labyrinthiform to sinuous, irregular, sometimes poroid, angular, absent near margin, 1-2 per mm; dissepiments up to 225 μm thick; pore tubes concolorous with the pore surface, sunk into equal depths, 3-4 mm long. Context white with a greyish or brownish tinge, often with a thin brown layer, less than 0.5 mm thick.

Hyphal system monomitic. Generative hyphae hyaline, thin to thick-walled branched, septate with clamps, sometimes breaking at nodes, 3-5 μm in diam. Cystidia basidia and basidiospores not observed.

DECAY : Causing white rot, mostly confined to perepheral region, white mycelia mass present in wood crevices.

SPECIMEN EXAMINED : On dead unidentified logs, from a semi-evergreen forest, Kuppadi, near Sultan Battery, Wyanaad Dist., Herb CALI No. G193, 19-6-1985.

GEOGRAPHICAL DISTRIBUTION : North and South America, Canada, Britain, France, USSR Australia, New Zealand.

REMARKS : The earlier Indian descriptions of this polypore were based on collections from the Himalayan regions (Thind & Rattan, 1971b; Bakshi, 1971). A widely known polypore, *A. biennis* has been described from various countries by Fries (1874), Rea (1922), Bourdot & Galzin (1927), Lowe (1942), Overholts (1953), Bonderstev (1953), Cunningham (1965), Ryvarden (1976a), Ryvarden & Johansen (1980) and Gilbertson & Ryvarden (1986). Development of a reddish colour on the pore surface on bruising reported by Lowe (1942) and Cunningham (1965) was not observed in the present collection.

AMYLOSPORUS Ryv.

Norw. J. Bot. 20 : 1, 1973.

Fruitbody terrestrial or lignicolous, stipitate or sessile. Pileus surface ochraceous to buff. Pores small and entire. Context white to pale brown. Hyphal system dimitic; generative hyphae hyaline, thin-walled, either simple-septate or with clamps often on the same hyphae; skeletal hyphae hyaline, thick-walled to almost solid, unbranched or moderately branched; gloeopleurous hyphae present, often difficult to observe. Cystidia absent. Basidiospores blue to grey in Melzer's, ellipsoid, smooth or finely warted.

Type species : *Tyromyces graminicola* Murr. [=*Amylosporus compbelllii* (Berk.) Ryv.]

Only one species has been collected from Kerala.

AMYLOSPORUS CAMPBELLII (Berk.) Ryv.

Norw. J. Bot. 24 : 217, 1977.

Pl. 21, fig. b-d.

Polyporus cambellii Berk., Hook. J. Bot. 6 : 228, 1854.

Tyromyces graminicola Murr., Trop. Polyp. P. 21, 1915.

Polyporus friabilis Bose, J. India Bot. 2 : 300, 1922.

P. propinquus Lloyd, Mycol. Writ. 7 : 1109, 1922.

Scutiger tisdalei Murr., Lloydia 6 : 227, 1943.

Fruibody annual, solitary to imbricate, epixylous but growing close to the soil surface, attached with a lateral stipe; flabelliform to spathulate, often lobate towards the margin, applanate, thick, fleshy while fresh, friable when dry; 6-9. 5 × 5-7 × 0. 5-0. 8 cm. Pileus surface greyish white (1B1) to brownish grey (5C2) or white with large pinkish patches, white towards margin; rough and uneven, finely velutinate with hairs agglutinated to form small pin-head like structures; margin smooth, wavy, incurved when dry. Stipe lateral, concolorous with the pileus surface or with brownish red (10D6) to greyish Magenta (13D4) patches, colour fading to yellowish brown (5E8) when dry; almost rectangular, up to 6cm long and 1.5 cm broad. Pore surface white with a brownish tinge even, and smooth; pores extending almost to the margin, visible to naked eye, never decurrent, round to angular, 4-5 per mm; pore mouth 110-200 μm wide; dissepiments 35-50(65) μm thick; pore tubes concolorous with the pore surface, 2-3 mm long, sunk into equal depths. Context white, soft and fragile, up to 4 mm thick, homogeneous, of loose intertwined hyphae. Whole sporophore becoming yellowish brown (5E8) when dry.

Hyphal system dimitic. Generative hyphae hyaline, thin-walled, septate with clamps, 3-5.6 μm in diam. Skeletal hyphae hyaline, thick-walled with narrow but distinct lumen, non-spetate, sparsely branched, 3.75-7.5 μm in diam. Cystidia absent. Basidia not observed. Basidiospores hyaline, ellipsoid to ovoid, thin-walled, smooth, amyloid, 4-5 × 3 3.5 μm.

DECAY : A brown cuboidal rot.

SPECIMEN EXAMINED : On logs of *Tectona grandis* Linn. (VERBENACEAE), lying in a semi-deciduous forest, Kanjirakadavu, Nilambur forests, Malappuram Dist., Herb. CALI No. G20, 19-7-1983.

GEOGRAPHICAL DISTRIBUTION : North America, Brazil, Venezuela, Jamaica, Bermuda, East Africa, India.

REMARKS : This polypore was described by Bose (1921b) from India as *Polyporus friabilis.* Bakshi's (1971) description of this polypore shows terrestrial and mesopodial fruitbodies with variable pileus surface but the micro characters are similar to those of the present collection. The East African (Ryvarden & Johansen, 1980) and North American (Gilbertson & Ryvarden, 1986) collections of *A. compbellii* have smaller pores and abundant gloeopleurous hyphae.

BJERKANDERA Karst.

Medd. Soc. Fauna Fl. Fenn. 5 : 38, 1879.

Fruitbody pileate, annual, soft and pliable. Pileus surface finely tomentose to smooth. Pore surface grey to black or buff to greyish brown, tubes concolorous. Context white to buff, black line present or absent between tubes and context. Hyphal system monomitic; generative hyphae with clamps, thin to thick-walled. Cystidia absent. Basidiospores smooth, thin-walled and nonamyloid. On angiosperms, very rare in conifers. Decay, a white rot. Cosmopolitan genus earlier known with two species.

Type species : *Bjerkandera adusta* (Fr.) Karst.

Only one species has been recorded from Kerala.

BJERKANDERA sp.

Fig. 14. e-g; Pl. ; Fig.

Fruitbody epixylous, annual, sessile, initially effused-reflexed, later becoming pileate, attached with a broad lateral base but mycelial haustoria arising from its centre alone, highly imbricate and confluent, dimidiate, applanate, corky while young, tough whey dry; 2.5-6 × 1.54 × 0.5-2.5 cm. Pileus surface dirty white to yellowish grey (3B2), white or light grey (1B1) near margin, colour darkening on handling; mostly uneven, faintly striate towards margin, azonate elsewhere, radially wrinkled, glabrous; margin smooth, even, becoming incurved on drying. Pore surface dull white with a creamy tinge, even and smooth; pores small and almost visible to naked eye, not extending to the margin, round, often irregular when two or more pores fusing to form a bigger pore, 5-7 per mm; pore mouth 150-180(700) µm thick; pore tubes concolorous with the pore surface, sunk into equal depths, 2-4 mm long, straight. Context white, floccose, homogeneous, sometimes considerably thinner towards margin, 0.6-2 cm thick.

Hyphal system monomitic. Generative hyphae hyaline, thin-to thick-walled, septate with frequent clamps, thick-walled hyphae mostly restricted to the upper context, 2-3(6) µm in diam. Cystidia none. Basidia clavate, 2 or 4-spored, 10-13 × 4.5-6 µm. Basidiospores hyaline cylindrical, smooth, thin-walled, 1 to 2-guttulate, nonamyloid, 5-6 × 2-3 µm; spore print white.

DECAY : Associated with white rot.

SPECIMEN EXAMINED : On decaying trunks of *Cocos nucifera* Linn. (PALMAE), Botanical Garden, Calicut University Campus, Malappuram Dist., Herb. CALI No. G26, 1-7-1984, G321, 25-6-1986, G379, 5-8-1987; on decaying logs of *Anacardium occidentale* Linn. (ANCARDIACEAE), Calicut University Campus, Malappuram Dist., Herb. CALI No. G29, 29-7-1983, G187, 13-6-1985.

REMARKS : The present collection is close to *Polyporus osteriformis* Berk. Reported from India by Bakshi (1971) and Roy (1969). Ryvarden (1984) considered *P. osteriformis* as synonym of *Bjerkandera fumosa* (Fr.) Karst. There are two species of *Bjerkandera, B. fumosa* having a dark layer at the base of the tubes and *B. adusta* (Willd. ex Fr.) Karst. lacking the black layer at the base of tubes. The present collection differs from both these species in certain respects. It lacks a dark layer at the base of the tubes which is specific for *B. fumosa* and differs from *B. adusta* in the colour of the pore surface and much longer basidia. It is likely that this is a new species.

CORIOLOPSIS Murr.

Bull. Torrey Bot. Club 32 : 358, 1905.

Fruitbody annual, pileate and sessile, rarely resupinate. Pileus surface yellowish to umber brown, often greysh, velutinate to hirsute, more rarely glabrous, zonate or azonate. Pore surface yellowish to umber brown, in some species with a greyish bloom; pores round and entire, small to moderately large. Context golden to umber brown. Hyphal system trimitic. generative hyphae hyaline, thin-walled with clamps

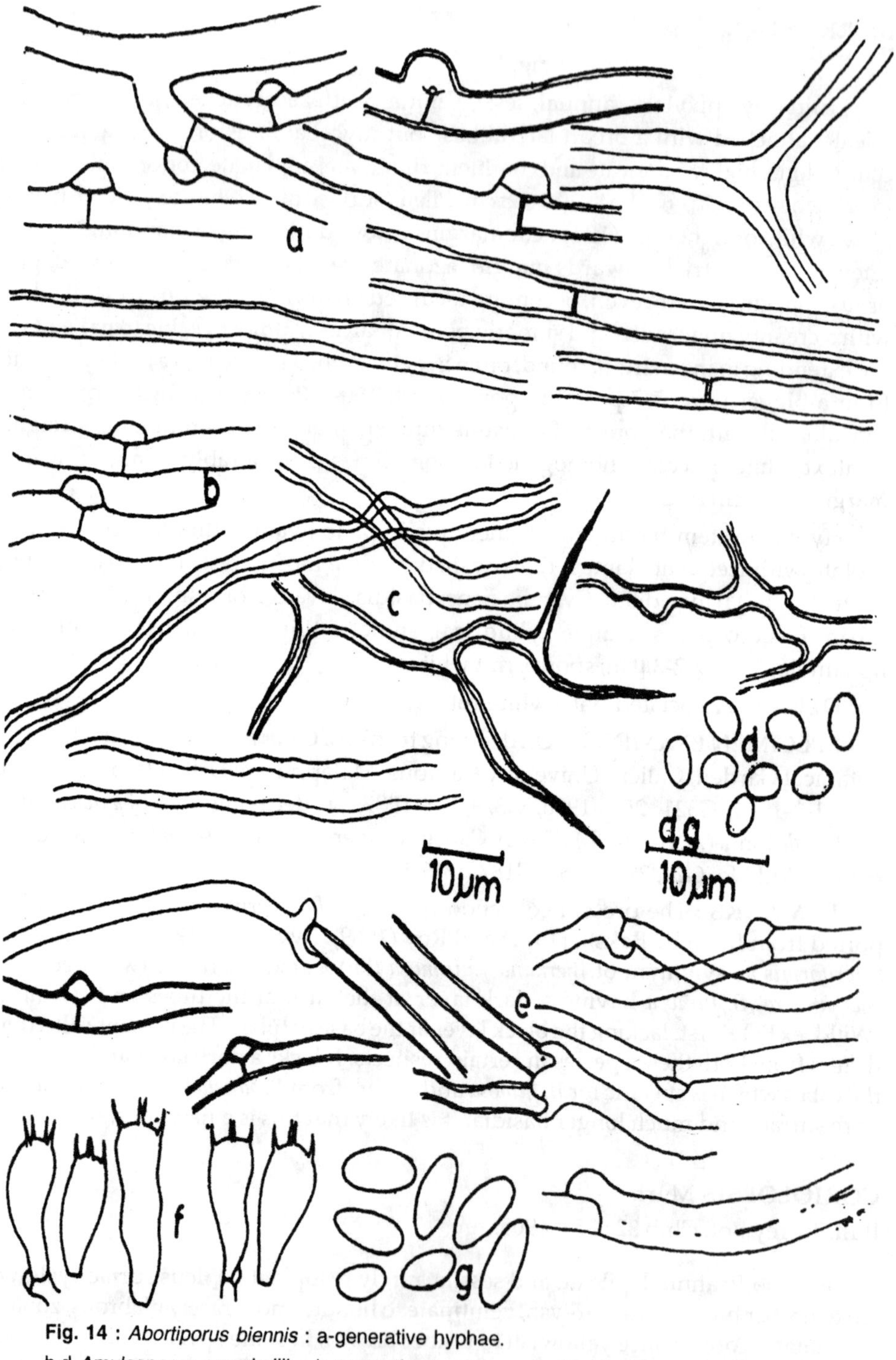

Fig. 14 : *Abortiporus biennis* : a-generative hyphae.
b-d *Amylosporus compbellii* : b-generative hyphae; c-skeletal hyphae; d-basidiospores.
e-g *Bjerkandera* taxonomic species : e-generative hyphae; f-basidia; g-basidiospores.

at the septa; binding and skeletal hyphae ochraceous to deep golden brown, thick-walled to solid. Cystidia absent but vegetative hyphae may project into the hymenium. Basidiospores hyaline, oblong-ellipsoid, smooth, thin-walled and nonamyloid, 6-12 µm long. On deciduous wood. Cosmopolitan genus, but most species restricted to the tropical zone.

Type species : *Coriolopsis occidentalis* (Kl.) Murr. [=*C. Polyzona* (Pers.) Ryv.]

Three species have been recorded from Kerala.

KEY TO SPECIES

1. Pileus surface glabrous; pores 6-8 per mm; dissepiments up to 50 µm thick.......... *C. sanguinaria*

1′. Pileus surface hirsute to scrupose; pores less than 5 per mm; dissepiments more than 50 µm thick.......... 2

2. Pileus surface yellowish, hispid to scrupose; pores angular or alveolar, up to 2 per mm; dissepiments often project as sharp irpicoid structures.......... *C. telfarii*

2′. Pileus surface browish, soft tomentose to hirsute; pores round, 3-5 per mm; dissepiments regular and smooth.......... *C. caperata*

CORIOLOPSIS CAPERATA (Berk.) Murr.

North Am. Fl. 9 : 77, 1908.

fig. 15. a-c Pl. fig.

Polyporus caperatus Berk., Ann. Nat. Hist. 1, 3 : 391, 1839.

P. phocinus Berk. & Br., J. Linn. Soc. 14 : 52, 1873.

Trametes dibapha Berk., Warm, Vidensk. Meddel. 31-32, 1880.

Polyporus purpureo-badius Pat., Bull. Soc. Mycol. Fr. 8 : 53, 1892.

Polystictus ekundiensis Henn., Engl. Bot. Jahrb. 22 : 91, 1897.

P. fisheri Henn., Ibid. 23 : 546, 1897.

P. griseo-brunneus Henn. ex Sacc. & Syd., Syll. Fung. 14 : 187, 1899.

Coriolopsis subglabrescens Murr., North Am. Fl. 9 : 77, 1908.

Fruitbody annual or reviving during the subsequent season, effused-reflexed or numerous imbricate fruitbodies connate to form a complex pileus, pileate fruitbodies sometimes attached with a central base, sub-stipitate or sessile, semi-circular to dimidiate, coriaceous when fresh, flexible when dry; 1.5-6 × 1.5-4.5 × 0.1-0.3 cm. Pileus surface mustard brown (5E6) to tobacco brown (5F6), sometimes hair brown (5E4), margin silver grey (4E2) to greyish white (IBI); soft tomentose near the margin to hirsute towards the base, almost glabrous when old, concentrically zonate and striate, shiny; margin thin and entire. Stipe when present concolorous with the pileus, up to 3mm long, 5 mm thick, discoid at the base, solid. Pore surface wax white (2B2) to silver white (2B2) or yellowish white (3B2), older specimens champagne (4B4), mostly even and smooth; pores small but visible to naked eye, absent near margin, round to oval or rarely elongated, irregular, 3-5 per mm; pore mouth 175-250 µm wide; dissepiments 50-125 µm thick; pore tubes concolorous with the pore surface, up to 1 mm long, often oblique, sunk into equal depths. Context concolorous with the pileus surface, dark brown in KOH.

Hyphal system trimitic. Generative hyphae hyaline, thin-walled, branched, rarely septate with clamps, up to 2.5 µm in diam. Binding hyphae hyaline to light brown, slightly thick-walled with distinct lumen, moderately branched, simple-septate, zig-zag., 2.5-3.75 µm in diam. Skeletal hyphae brown, thick-walled with broad lumen, non-septate, unbranched, 3.75-5 µm in diam. Hyphal pegs present. Basidia and basidiospores not observed.

DECAY : A white fibrous rot.

SPECIMEN EXAMINED : On dead forest logs, host not identified, Kaduvakkunnu, Nadugani forest, Nilambur, Malappuram Dist., Herb. CALI No. G52, 28-2-1984, G129, 13-11-1984; Silent Valley National Park, Palghat Dist., Herb. CALI No. G57, 30-5-1984; on forest logs, Kanoth Reserve Forest, Wyanaad Dist., Herb. FRI No. 4373, K. Bagchee, 6-4-1946; Boonthakkal, Begur Reserve Forest, Wyanaad Dist., Herb. FRI No. 5150, K. Bagchee, 23-1-1950.

GEOGRAPHICAL DISTRIBUTION : India, Sri Lanka, Philippines, Africa, Mauritius, America, Brazil, Mexico, French Antilles and French Guiana.

REMARKS : This polypore is already reported from North India by Bose (1937a) and Bakshi (1971). According to Bakshi (1971) *Polyporus caperatus* has dimitic hyphal system and clampless generative hyphae. All the fruitbodies collectecd from India so far were sterile.

CORIOLOPSIS SANGUINARIA (Kl.) Teng.

True fungi of China p. 760, 1964.

fig. 15. d-h Pl. fig.

Polyporus sanguinarius Kl., Linnaea 8 : 484, 1833.

P. pruinatus Kl., Ibid. 8 : 486, 1833.

P. bicolor Jungh., Verh. Batav. Genootsch. 17 : 54, 1838.

P. anebus Berk., Land. J. Bot. 6 : 504, 1847.

P. loreus Beeli, Bull. Jard, Bot. Etat. Brux. 8 : 225, 1930.

Fruitbody annual, resupinate initially, becoming effused-reflexed or even pileate, if pileate then sessile, attached with a converging lateral base, densely imbricate, confluent, flabelliform to petaloid to reniform, rarely semi-circular, applanate, thin, coriaceous while fesh, rigid when dry; 2-4.5 × 1.5-2.7 × 0.15-0.2(0.4) cm. Pileus surface greyish orange (5B4) to plae orange (5A3), golden blonde (7E5) towards margin; radially striate, striations faint, concentrically zonate, zonations more prominent towards margin in some fruitbodies, faintly crusty, glabrous, reflexed when dry, context hyphae arising in irregular blocks on the pileus surface; margin entire, smooth and rigid. Pore surface greyish orange (5B4) when young, becoming ivory (4B4) to blonde (4C4) or greyish orange (5B3) when mature, smooth and even; pores extending to the margin, minute and not visible to the naked eye, round to somewhat angular, regular, 6-8 per mm; pore mouth 75-90(120) µm wide; dissepiments (25) 35-50 µm thick; pore tubes concolorous with the pore surface or raw Sienna (6D7), 1-1.5 mm long, sunk into equal depths. Context orange white (5A2) or pale yellow (4A3), up to 2mm thick.

Hyphal system trimitic. Generative hyphae hyaline, thin-walled, branched, nodoseseptate, 2-3 µm in diam. Binding hyphae hyaline to pale-coloured, thick-walled, with distinct lumen, highly branched, non-septate, 2.5-3.75(5) µm in diam. Skeletal hypahe light yellow, thick-walled with distinct lumen, unbranched or with short terminal branches, non-septate, cylindrical and equal in width, (4.5)6.25-7.5 µm in diam. Cystidia none. Basidia clavate, 2 or 4-spored, 10.5-15 × 4.5-6 µm; sterigmata up to 1.5 µm long. Basidiospores hyaline, oblong ellipsiod to sub-cylindrical, smooth, nonamyloid 5-7.5 × 2-3 µm.

DECAY : White rot.

SPECIMEN EXAMINED : On unidentified forest logs, Kaduvakkunnu, Nadugani forest, Nilambur, Malappuram Dist., Herb. CALI No. G137, 13-11-1984; on fallen tree branches, Silent Valley National Park, Palghat Dist., Herb. CALI No. G225, 15-9-1985.

GEOGRAPHICAL DISTRIBUTION : Paleotropical species; Africa, Mauritius, India, China, Philippines.

REMARKS : Bakshi (1971) reported this polypore from India as *Polyporus bicolor.* The present collection agrees with the Chinese collection in its macro-and micromorphology (Zhao Ji Ding, 1986 : personal communication). The East African sporophores (Ryvarden & Johansen, 1980) are much thicker than the present collection while there is agreement with the description of *C. Sanguinaria* by Ryvarden (1979b) and Lloyd (1915b).

CORIOLOPSIS TELFARII (Kl.) Ryv.

Norw. J. Bot. 19 : 230, 1972.

fig. 15. i-l; Pl. fig.

Polyporus telfarii Kl., Linnaea 8 : 484, 1833.

P. zeylanicus Berk., Ann. Nat. Hist. 10 : 377, 1843.

Trametes cristata Cke., Grevillea 10 : 132, 1886.

Hexagonia dybowskii Pat., Bull. Soc. Mycol. Fr. 8 : 54, 1892.

Trametes wildermaii Bres., Ann. Mycol. 9 : 269, 1911.

Fruitbody annual, pileate and sessile, resupinate to effused-reflexed, variously connate, pileate forms attached with a narrow lateral base or with a rudimentary stipe, coriaceous while fresh, semi-rigid when dry; effused forms up to 20 × 15 × 0.3

cm, pileate form 4-9X 3-4.5 × 0.2-0.5 cm. Pileus surface pastel yellow (3A4) champagne (4B4) or butter yellow (4A5), lighter towards margin, hispid to scrupose, soft to touch, concentrically striate or zonate, radially wrinkled, hairs more prominent on the upper ridges; margin lighter, wavy, irregularly spinose, rigid, hairs almost absent. Stipe very small, up to 1 mm long, concolorous withthe pileus surface, solid. Pore surface golden blonde (5C4) to clay (5D5), margin slightly greyish orange (5B4), uneven, rough; pores large, arranged in somewhat radial rows, absent near the margin, angular to alveolar or irregular, round and regular towards margin, -1-2 per mm; dissepiments project as irpicoid and rigid structures, 100-150 μm thick; pore tubes yellowish or lighter than pore surface, inner surface sometimes greyish 1-3 mm long, sunk into unequal depths. Context concolourous with pore tubes, up to 3mm thick, fibrous, homogeneous.

Hyphal system trimitic. Generative hyphae hyaline, thin-walled, branched, nodose-septate, common in dissepiments and margin, 2-3 μm in diam. Binding hyphae hyaline to pale-coloured, well branched, non-septate, solid, 2-3(5) μm in diam. Skeletal hyphae pale-coloured, thick-walled with narrow lumen, unbranched, non-septate, main stem cylindrical and equal in width, 3.75-6.5 μm in diam. Cystidia none. Basidioles clavate, thin-walled, 15-20 × 5-7 μm. Basidiospores not observed.

DECAY : White rot.

SPECIMEN EXAMINED : On decaying logs of *Tectona grandis* Linn. f. (VERBENACEAE), Divisional Forest Office Campus, Nilambur, Malappuram, Dist., Herb. CALI No. G147, 7-3-1985; on unidentified forest logs, Karadimala, Brahmagiri Ranges, Tirunelli, Wyanaad Dist., Herb. CALI No. G95, 2-10-1985; Kaduvakkunnu, Nadugani forests, Nilamabur, Malappuram Dist., Herb. CALI No. G347, 30-8-1986.

REMARKS : The present collections agree with the earlier Indian description of *P. zeylanicus* (Bakshi, 1971). Klotzsch's type specimen (Ryvarden, 1967b), Berkeley's collection (Ryvarden, 1976c) and East African collection (Ryvarden & Johansen, 1980) also agree with the present collection in their macro-and microcharacters.

FAVOLUS Fr.

Elench. Fung. 1 : 44, 1828.

Fruitbody annual, dimidiate, flabelliform, or spathulate, rarely with a centrally placed stipe, thin and soft when fresh, brittle and light-in-weight when dry. Pileus surface white, cream or straw-coloured but may darken with age, glabrous or slightly tomentose, smooth, tesselate to pustulate or with distinct thin radial lines or fine striate. Pores hexagonal to radially elongate, semi-lamellate towards stipe; dissepiments paper-thin, often undulating. Context very thin except close to the base. Hyphal system dimitic; generative hyphae with simple-septa; skeletal hyphae with weak branchings, thick-walled with out a distinct lumen. Cystidia absent. Basidiospores cylindrical to slightly navicular with tapering ends, smooth, thin-walled and negative in Melzer's. On angiosperms. Pantropical genus with two variable species.

Type species : *Favolus brasiliensis* (Fr.) Fr.

Only one species has been recorded from Kerala.

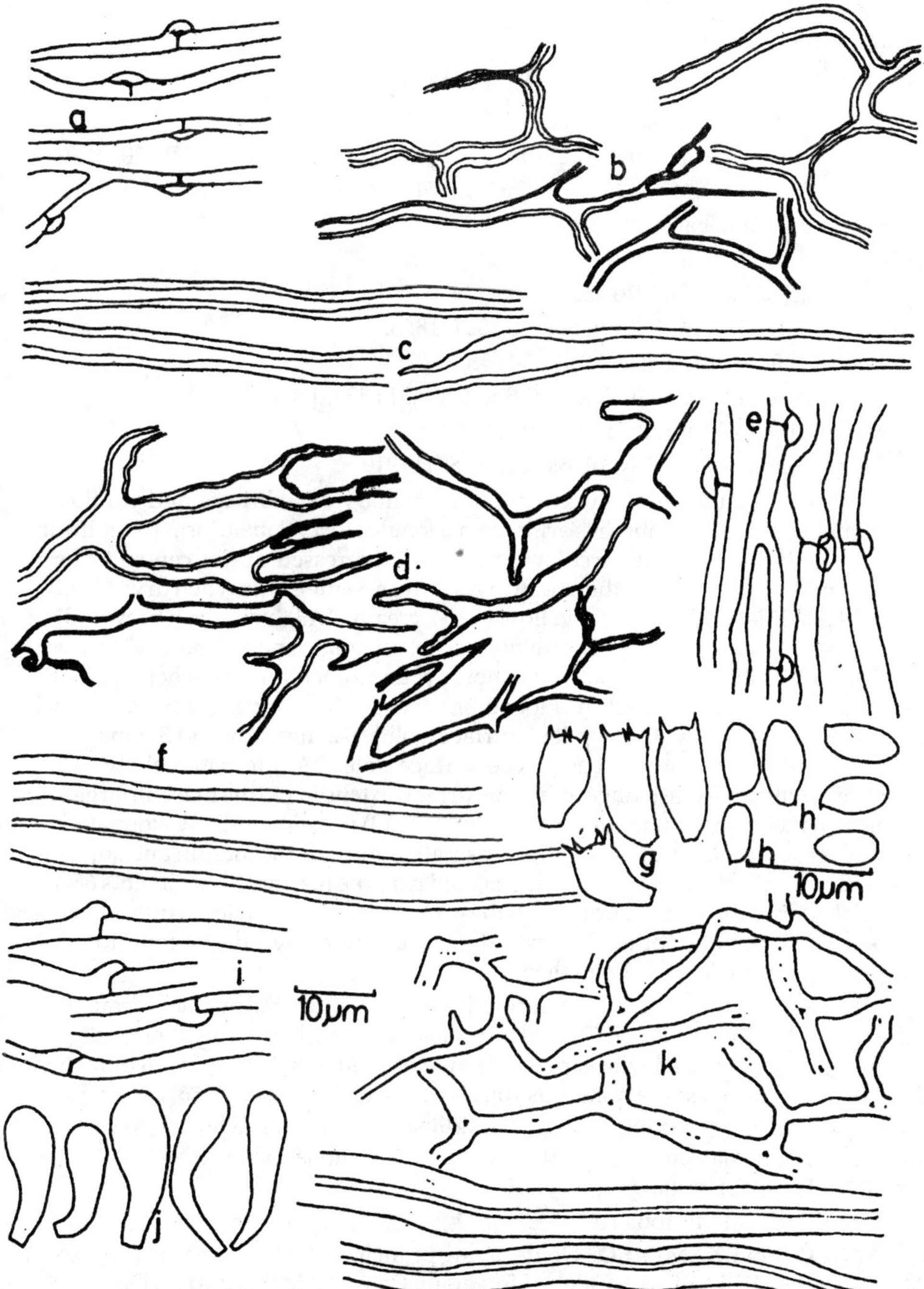

Fig. 15 : a-c *Coriolopsis caperata* : a-generative hyphae; b-binding hyphae; c-skeletal hyphae. d-h *Coriolopsis sanquinaria* : d-binding hyphae; e-generative hyphae; f-skeletal hyphae; g-basidia; h-basidiospores.

i-l *Coriolopsis telfarii* : i-generative hyphae; j-basidioles; k-binding hyphae; l-skeletal hyphae.

FAVOLUS BRASILIENSIS (Fr.) Fr.

Elench. Fung. 1 : 44, 1828,

fig. 16. a-d; Pl. fig

Daedalea brasiliensis fr., Syst. Mycol. 1 : 332, 1821.

Polyporus dermoporus Pers. In Gaudichaud, Voyage aut. Monde, Paris. P. 170,1827.

Favolus quadalupensis Lev., Ann. Sci. Nat. 3,5 : 511, 1830.

F. hepaticus Kl. Linnaea 7 : 197, 1832.

F. alutaceus Berk. & Mont., Ann. Soc. 11 : 240, 1849.

F. friesii Berk. & Curt. Ibid. 10 : 321, 1868.

F. hispulus Berk., & Curt. J. Linn. Soc. 10 : 321, 1868.

F. bifindensis Henn., Engl. Bot. Jahrb. 30 : 43, 1902.

Hexagonia wilsonii Murr., Bull. Torrey Bot. Club 31 : 329, 1904.

H. daedales Link ex Murr., Ibid 31 : 328, 1904.

Favolus caespitosus Lloyd, Mycol. Notes 58 : 821, 1919.

Fruitbody annual, laterally sub-stipitate, solitary or imbricate and confluent, sometimes a group of fruibodies arise from a single stock; flabelliform to reniform, seldom spathulate or semi-circular, sometimes depressed at the centre to form semiinfundibuliform pileus, thin, fleshy while fresh, semi-brittle when dry; 3-8 X 3-8 X 0.2-0.4 cm. Pileus surface light yellow (4A4), cream (4A3), yellowish brown (5E8) towards margin, azonate, faintly radially striate, pustulate and tesselate while fresh, basal portion of the pileus surface with short hyaline tomentum, elsewhere glabrous; margin wavy, smooth, sometimes lobed, incurved when dry. Stipe usually much reduced, concolorous with the pileus surface, solid, 1-2 mm long and 3-4 mm thick attached with a thin discoid base. Pore surface pale (2A2) to pale yellow (3A3), sometimes with a greenish tinge; pores decurrent, extending to the margin, large and visible to naked eye, pentagonal to hexagonal, 1-2(3) per mm, radially elongated up to 2 mm, pores in radiating rows, slightly smaller towards the margin and stipe (up to 3 per mm), sometimes becoming daedaleoid near the margin; dissepiments 65-135 µm thick; pore tubes concolorous with the pore surface, 2-3 mm long, tubes straight and sunk into equal depths. Context white to cream coloured, thinning towards margin, homogeneous, 0.5-2 mm thick.

Hyphal system dimitic. Generative hyphae hyaline, thin-walled, branched, simple septate, uncommon, 1.75-2.5 µm in diam. Binding hyphae hyaline to slightly pale yellow, thick-walled with capillary lumen or solid, non-septate, well branched, 2.5-6 µm in diam. Cystidia none. Basidia clavate, 2 or 4-spored, 13-16 × 4.5-5.5 µm; sterigmata up to 3 µm long, straight. Basidiospores hyaline, cylindric to clindric-ellipsoid, slightly naviculate, smooth, 1-to 3-guttulate, nonamyloid, 5-8 × 2.3-3.5 µm; spore print white. Hyphal pegs present.

DECAY :White fibrous rot.

SPECIMEN EXAMINED : On decaying branches of *Macaranga peltata* (Roxb.) Muel. (EUPHORBIACEAE), Calicut University Campus, Malappuram, Dist., Herb. CALI No. G324, 6-71986; on decaying unknown forest logs, pathanthode, Silent Valley National Park , Palghat Dist., Hexb CALI No. G32, G100, 6-91983 ; Kanjirakadavu, Nilambur Forest Division, Malappuram, Dist., Herb. CALI No. G102, 30-6-1983; Mundakal forests, Meppadi, Wayanaad Dist., Herb CALI No. G117 , 22-9-

1984; Kaduvakkunnu, Nadugani forests, Nilambur, Malappuram Dist., Herb. CALI No. G131, 13-11-1984; Silent Valley National Park, Palghat Dist., Herb. CALI No. G233, 11-9-1985; Perumpara (W), Sholayar forests, Trichur Dist., Herb. CALI No. G339, 30-8-1986.

GEOGRAPHICAL DISTRIBUTION : North America, East Africa, Mexico, French Antilles, French Guiana, India.

REMARKS :The present collection appear similar to the earlier Indian collection of *F. brasiliensis* (Bose, 1937a; Bakshi, 1971). Bakshi (1971), however, describes a monomitic hyphal system for this species. The North American collections (Overholts, 1953) have bigger spores while the east African fruitbodies (Ryvarden & Jonansen, 1980) have broader hyphae and bigger spores.

FLAVODON Ryv.

Norw. J. Bot. 20 : 3, 1973

Fruitbody annual, resupinate to pileate, whole fruitbody becoming reddish to brown with KOH. Pilues suface yellowish to ocharaceous-grey, abpressed-tomentose. Hymenophore first poroid, then hydnoid to irpicoid, Context bright sulphurous yellow. Hyphal system dimitic; generative hyphae simple septate; skeletal hyphae thick-walled and partly bent into the hymenium as smooth or cncrusted cystidia. Basidiospores broadly ellipsiod, smooth, hyaline and nonamyloid. On deciduous wood. Monotypic genus.

Type species : *Flavodon flavus* (Kl.) Ryv.

FLAVODON FLAVUS (Kl.) Ryv.

Norw. J. Bot. 20 : 3, 1973

fig. 16. e-i Pl. fig.

Irpex flavus Kl., Linnaea 8 : 488, 1833.
Polyporus flavus Jungh., Prae. Fl. Crypt. Javae Ins. P. 46, 1838.
Irpex flavus Jungh., Varh. Batav. Genootsch 17 : 46, 1839, nomen illegit..
Polystictus flavus (Jungh.) Fr., Nova Acta 3, 1 : 85, 1851.
Polyporus crenatoporus Rostr., Bot. Tidsskar. 24 : 360, 1902.
Coriolus melleoflavus Murr., Bull. Torrey Bot.Club 35 : 393, 1908.
Poria crocea Lloyd, Mycol, Notes 66 : 1129, 1922, nomen nudum.
Hydnum luteomarginatum Beeli, Bull. Soc. Roy. Bot. Belg. 58 : 210, 1926.
Trichaptum flavum (Jungh.) G.H. Cunn., N.Z. Dep. Sci. Industr. Res. Bull. 164 : 101, 1965.

Fruitbody annual, effused-reflexed, rarely pileate, when pileate attached with a broad lateral base, usually with numerous small imbricate pilei, variously connate; semicircular and applanate, corky to coriaceous in consistency while fresh as well as dry; 3.5-5 × 2-4 × 0.1-0.3 (0.5) cm. Pileus surface wine yellow (3B3) to alabaster (5B2), Margin pale yellow (3A3); adpressed-tomentose, tomentum sometimes agglutinated to appear rigid and scrupose, often concentrically zonate and ridged; margin entire, thin often hairy. Pore surface lemon yellow (3B8) while young, ash blonde (3C3) or slightly darker when mature; uneven, irpicoid to hydnoid, vaguely poroid near margin in some sporophores; spines citrine yellow while young becoming brown to ochraceous with age, cylindrical to somewhat subulate, ends flattened, irregularly distributed, absent near margin, up to 2 per mm, up to 3 mm long, context lemon yellow (3B8), up to 2 mm thick, fibrous, of compactly arranged lower stratum with horizontal hyphae and loosely arranged upper stratum with vertically oriented hyphae. Whole fruitbody becoming reddish in KOH.

Hyphal system dimitic. Generative hyphae hyaline, thin-walled, branched, simple septate, 2.5-3.75 µm in diam. Skeletal hyphae, 3.75-6.5 µm in diam. Skeletocystidia present; skeletal hyphae protruding into the hymenium with apical encrustations, 6-7 µm broad. Basidia long, narrowly clavate, 17.5-32.5 × 5 µm. Basidiospores hyaline, ovoid to broadly ellipsoid, thin-walled, nonamyloid, smooth, 5-5.25 × 2.5-3 µm.

DECAY : White fibrous rot.

SPECIMEN EXAMINED : On dead stumps of *Anacardium occidentale* Linn. (ANACARDIACEAE), near Calicut University Campus, Malappuram Dist., Herb. CALI No. G281, 2-4-1986; on decaying branches of *Delonix regia* (Hook.) Rafin. (PALILIONACEAE). Calicut University Campus, Malappuram Dist., Herb. CALI No. G69, 27-11-1983; on branchlets of *Tectona grandis* Linn. F. (VERBENACEAE), Kerala Forest Research Institute Sub-Centre, Nilambur, Malappuram Dist., Herb. CALI No. G25, 19-7-1983; on unknown host, Wyanaad forests, Wyanaad Dist., Herb. FRI No. 4429, K. Bagchee, 13-8-1946.

GEOGRAPHICAL DISTRIBUTION : Tropical Africa, Philippines, Australia, New Zealand, Pakistan, India.

REMARKS : Bakshi's (1971) description of *Irpex flavus* shows slightly narrower cystidia and spores while Roy (1981a) reports thinner skeletal hyphae, wider basidia and longer spores than those of the present collection. Except for their bigger spores both the Australian (Cunningham, 1965) and East African (Ryvarden & Johansen, 1980) collections of this polypore agrees with the present collection. A non host-specific polypore, *F. Flavus* is common throughout Kerala. The fungus sporulates well during July-August,

FOMITOPSIS Karst.

Medd. Soc. Fauna Fl. Fenn. 6 : 9, 1881.

Fruitbody perennial or rarely annual, sessile to effused-reflexed, tough and woody hard, often large. Pileus surface smooth, usually sulcate, with distinct crust. Pore surface white, light cream or pink, pores mostly small, regular, round, tubes concolorous with pore surface, stratified. Context white, cream or pink, suberose and tough. Hyphal system trimitic; generative hyphae with clamps; skeletal hyphae and binding hyphae hyaline. Cystidia none. Basidiospores hyaline, cylindrical to ellipsoid, smooth, thin-walled, nonamyloid. On dead conifers and deciduous wood. Cosmopolitan genus.

Type species : *Fomitopsis pinicola* (Swartz ex Fr.) Karst.

Four species have been recorded from Kerala.

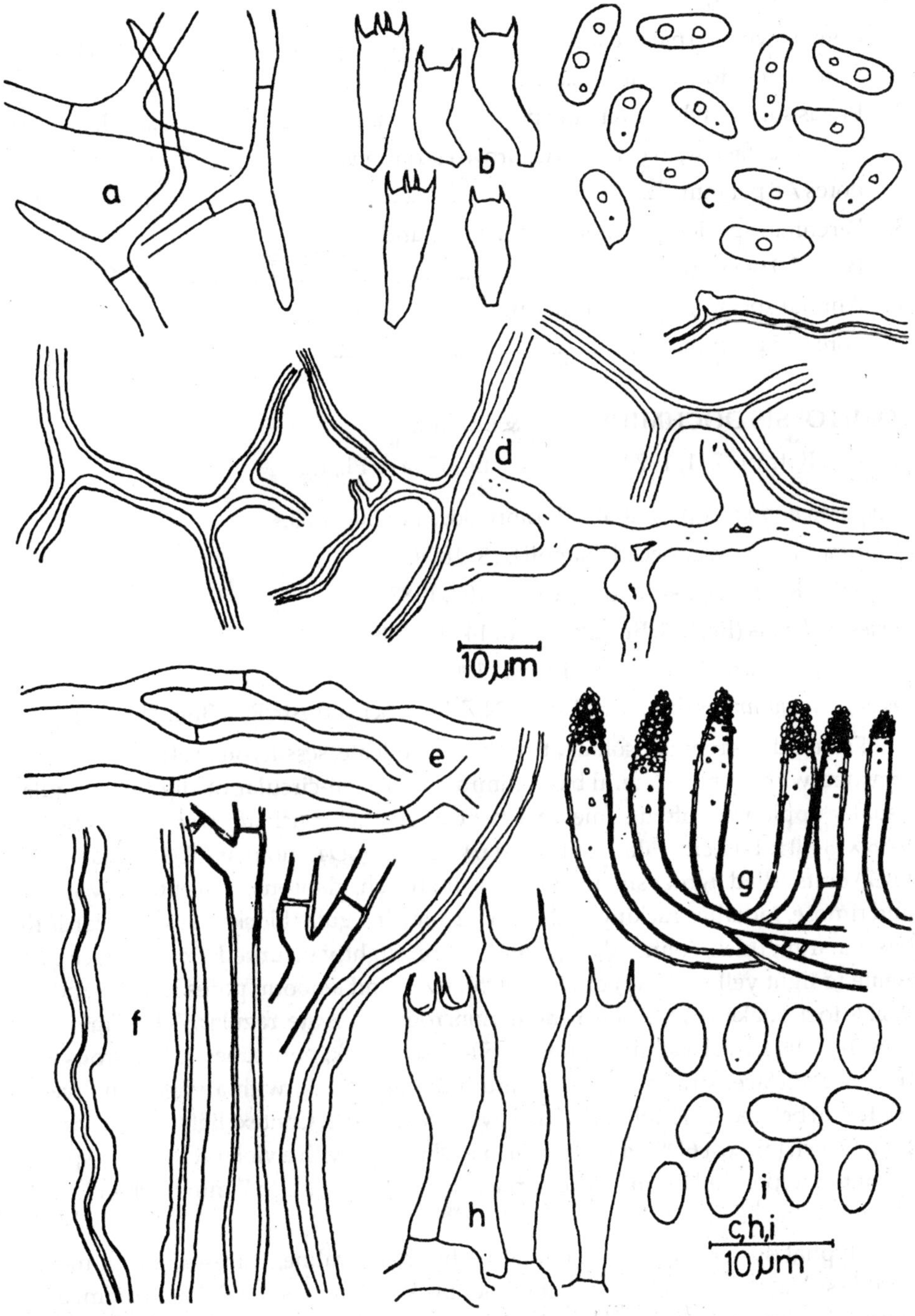

Fig. 16 : a-d *Favolus brasiliensis* : a-generative hyphae; b-basidia; c-basidiospores; d-binding hyphae.

e-i *Flavodon flavus* : e-generative hyphae; f-skeletal hyphae; g-skeletocystidia; h-basidia; i-basidiospores.

KEY TO SPECIES

1. Context white to pale yellow.......... 2
1′. Context rose to brownish.......... 3
2. Pileus surface white to cream; pores 4-7 per mm.......... *F. palustris*
2′. Pileus surface with concentric, brown striations; pores 7-8 per mm.......... *F. rhodophaeus*
3. Perennial species; pileus surface rimose and sulcate; Pores 7-9 per mm.......... *F. dochmius*
3′. Annual species; pileus surface not sulcate or rimose; pores 4-5 per mm.......... *F. feei*

FOMITOPSIS DOCHMIUS (Berk. & Br.) Ryv.

Norw. J. Bot. 19 : 231, 1972. fig. 17. a-c; Pl., fig.

Polyporus dochmius Berk & Br., J. Linn. Soc. 14 : 50, 1873.

P. aurora Ces., Roy.C. Accad. Napoli 8 : 5, 1879.

P. seriatus Kalch. ex Cke., Grevillea 10 : 102, 1882.

Fomes dochmius (Berk. & Br.) Cke., Ibid. 14 : 17, 1885.

F. subferreus Murr., North Am. Fl. 9 : 97, 1908.

Osmoporus dochmius (Br.), G.H. Cunn., N.Z. Dep. Sci. Industr. Res. Bull. 165 : 244, 1965.

Fruitbody perennial, solitary to imbricate, pileate, sessile, often effused-reflexed, attached with a broad lateral base, dimidiate to semi-circular or slightly irregular, usually applanate, often somewhat bent to appear conchate, woody and heavy; 9-88 × 8-40 × 1-17 cm. Pileus surface dark blonde (5D4), nougat (5D3), golden grey (4C2) or its allied lighter shades; uneven and rough, glabrous, concentrically sulcate and rimose, radially cracking and often forming irregular blocks, crusty towards the base; margin thick, rounded, rigid and entire, slightly incurved. Pore surface dirty white to light yellow (4A4) or clay (5D5), even and smooth; pores minute and not visible to the naked eye, absent near margin, round to ovate, regular, 7-9 per mm; pore mouth 75-90 μm broad; dissepiments 50-65 μm thick; pore tubes concolorous with the pore surface, stratified, each stratum 2-3 mm thick, with a very thin layer of context in between, old tubes invaded by context-tissue. Context light orange (5B4) to light brown (almost 6D8), dark brown in KOH, often with a violet stain, stain spreading into the pore tubes and pore surface also, fibrous, up to 7 mm thick below the crust, homogeneous.

Hyphal system trimitic. Generative hyphae hyaline, thin-walled, branched, simple septate or with clamps, sometimes broken at nodes, 1.5-2 μm in diam. Binding hyphae hyaline to pale-coloured, thick-walled, solid, branched, sometimes sparsely branched, non-septate, flexuous and twisted, 1.5-2.5 μm in diam. Skeletal hyphae pale-coloured to light brown, thick-walled, with narrow lumen, non-septate, usually unbranched, sometimes with short apical branches, cylindrical and straight, 2.5-5 μm in diam. Setae none. Basidia and basidiospores not observed.

DECAY : Brown rot, the wood broken into irregular blocks with white hyphal mats in the cracks.

SPECIMEN EXAMINED : On decaying trunk of *Mesua nagassarium* (Burm. F.) Kosterm. (CLUSIACEAE), Malakkapara, Sholayar forests, Trichur Dist., Herb. CALI No. G179, 10-6-1985, G356, 30-8-1986, Kanoth Range, Wyanaad Dist., Herb. FRI No. 5151, K. Bagchee, 22-1-1950, Chandanthody, Wyanaad Dist., Herb. FRI No. 4247, K. Bagchee, 24-1-1946; on decaying trunk of *Terminalia paniculata* Roth. (COMBRETACEAE), Perumpara (W), Malakkapara, Sholar forests, Trichur Dist., Herb. CALI No. G346, 30-8-1986; on decaying trunk of *Grewia tiliifolia* Vahl. (TILIACEAE), Perumpara, Malakkapara, Sholayar forests, Trichur Dist., Herb. CALI No. G344, 30-8-1986; on living trunk of *Dillenia pentagyna* Roxb. (DILLENIACEAE), Perumpara, Malakkapara, Sholayar forests, Trichur Dist., Herb. CALI No. G336, 30-8-1986; on unidentified forest trees, Panthenthode, Silent Valley National Park, Palghat Dist., Herb. CALI No. G37, 7-9-1983; Chooralmala, Meppadi, Wyanaad Dist., Herb. CALI No. G116, 21-9-1984; Poochipara, Silent Valley National Park, Palghat Dist., Herb. CALI No. G232, 15-9-1985; Parambiculam forests, Teak Plantation Division, Palghat Dist., Herb. CALI No. G273, 16-1-1986.

GEOGRAPHICAL DISTRIBUTION : North and South America, Sri Lanka, India, Philippines, Australia, New Guinea.

REMARKS : *F. dochmius* which is one of the most common perennial polypores of the evergreen and semi-deciduous forests of Kerala, causes severe brown cuboidal rot in commercially valuable timber. The density of its occurrence in forets is very high. Sporophores were collected from living trees as well as from decaying logs. All the collections showed sterile hymenium.

On comparing with earlier India reports, it is found that the sporophores described by Bose (1973a, 1946) have wider pore mouths and Bakshi (1971) described it as a dimitic species.

FOMITOPSIS FEEI (Fr.) Kreisel.

Univ. Habana Ser. 4, Cienc. Biol. 16 : 83, 1971.

fig. 27. d-h; Pl. fig.

Polyporus feei Fr., Linnaea 5 : 518, 1830.

P. rubidus Berk., Lond. J. Bot. 6 : 500, 1847.

Polyporus eucalypti Kalch. ex Thuem, Grevillea 4 : 73, 1875.

Polystictus feei (Fr.) Cke. Ibid. 14 : 82, 1886.

P. eucalypti (Kalch.) Cke. Ibid. 14 : 82, 1886.

Trametes feei (Fr.) Pat., Essai Tax. P. 92, 1900

Fomes feei (Fr.) Lowe, Mycologia 47 : 217, 1955.

Fruitbody annual, pileate, sessile, attached with a lateral converging base, solitary to imbricate, confluent, semi-circular to flabelliform, often dimidiate, applanate, corky in consistency while fresh, hard when dry; 1.5-7.5 × 0.2-0.5 cm. Pileus surface rust brown (7F6) towards the margin, young pileus ivory (4B4) to reddish blonde

(5C4); faintly pubescent while young soon becoming glabrous, smooth and even, sometimes slightly wavy, concentrically striate, radially faintly wrinkled, slightly umbonate near the attachment; margin rigid, smooth and entire. Pore surface ivory (4B3) while young, soon becoming somewhat reddish blonde (5C4), oak brown (5D6) when mature, slightly rough; pores small but visible to naked eye, absent near margin, round. sometimes irregular, walls often dissolving to form daedaleoid pores, 4-5 per mm; pore mouth 125-160 (225) μm wide; dissepiments 5— 75 μm thick; pore tubes concolorous with pore surface, sunk into equal depths, 1-2. 5 mm long, straight. Context sark brown (7F8) to agate (7E8), floccose, homogeneous, slightly shiny, up to 2 mm thick.

Hyphal system trimitic. Generative hyphae hyaline, thin-walled, branched, septate with clamps, sparsely present and easily deteriorating, 1. 5-2 μm in diam. Binding hyphae pale-coloured, arboriform, thick-walled to solid, non-septate, extremities attenuating, 2. 5-3. 5 μm in diam. Skeletal hyphae hyaline to light brown, unbranched or weakly branched at endings, thick-walled with narrow or obliterating lumen, sometimes solid, non-septate, either cylindrical and straight or flexuous and twisted, 3.75-5(6) μm in diam. Basidia narrowly clavate, 2 or 4-spored, 13.5-15 × 4.5-6 μm; sterigmata up to 2 μm long. Cystidia none. Basidiospores hyaline, thin-walled, cylindric-ellipsoid, apiculate, smooth, nonamyloid, 4-6 × 1.5-2.5 μm.

DECAY : Brown cuboidal rot with mycelial tufts in wood crevices.

SPECIMEN EXAMINED : On decaying logs of *Anacardium occidentale* Linn. (ANACARDIACEAE), Botanical Garden, Calicut University Campus, Malappuram Dist., Herb. CALI No. G70, 6-7-1984; on unidentified forest logs, Panthenthode, Silent Valley National Park, Palghat Dist., Herb. CALI No. G55, G63, 30-5-1984; Kanjirakadavu, Amarambalam Range, Nilambur Forest Division, Malappuram Dist., Herb. CALI No. G101, 30-6-1983.

GEOGRAPHICAL DISTRIBUTION : South and Central America, New Guinea, East Africa, Australia, New Zealand, Malaysia, Philippines, Thailand, Sarawak, Sri Lanka, India.

REMARKS : The present collection agrees well with the earlier Indian descriptions of the species (as *Polyporus ribidus*) (Bose, 1937a; Bakshi, 1971). The North Amarican (Lowe, 1957; Gilbertson & Ryvarden, 1986) and East African (Ryvarden & Johansen, 1980) sporophores have broader generative hyphae and bigger basidia.

FORMITOPSIS PALUSTRIS (Berk. & Cur.) Gilbn. & Ryv.

Mycotaxon 22 : 364, 1985. fig. 17. i-m Pl. fig.

Polyporus palustris Berk. & Curt., Hook. J. Bot. 1 : 102, 1849.

Tyromyces palustris (Berk. & Curt.) Ryv. Norw. J. Bot. 24 : 223, 1977.

Fruitbody annual to biennial, pileate, sessile, attached laterally with a broad base, imbricate and confluent, rarely solitary, semi-circular, applanate, leathery while fresh, rigid when dry; 4-8 X 3-5 X 0.51 cm. Pileus surface white to cream (4A3) often pale yellow (3A3); glabrous, concentrically striate, striations prominent while young, uneven and smooth when old, pelliculose, tuberculate towards base, incurved and

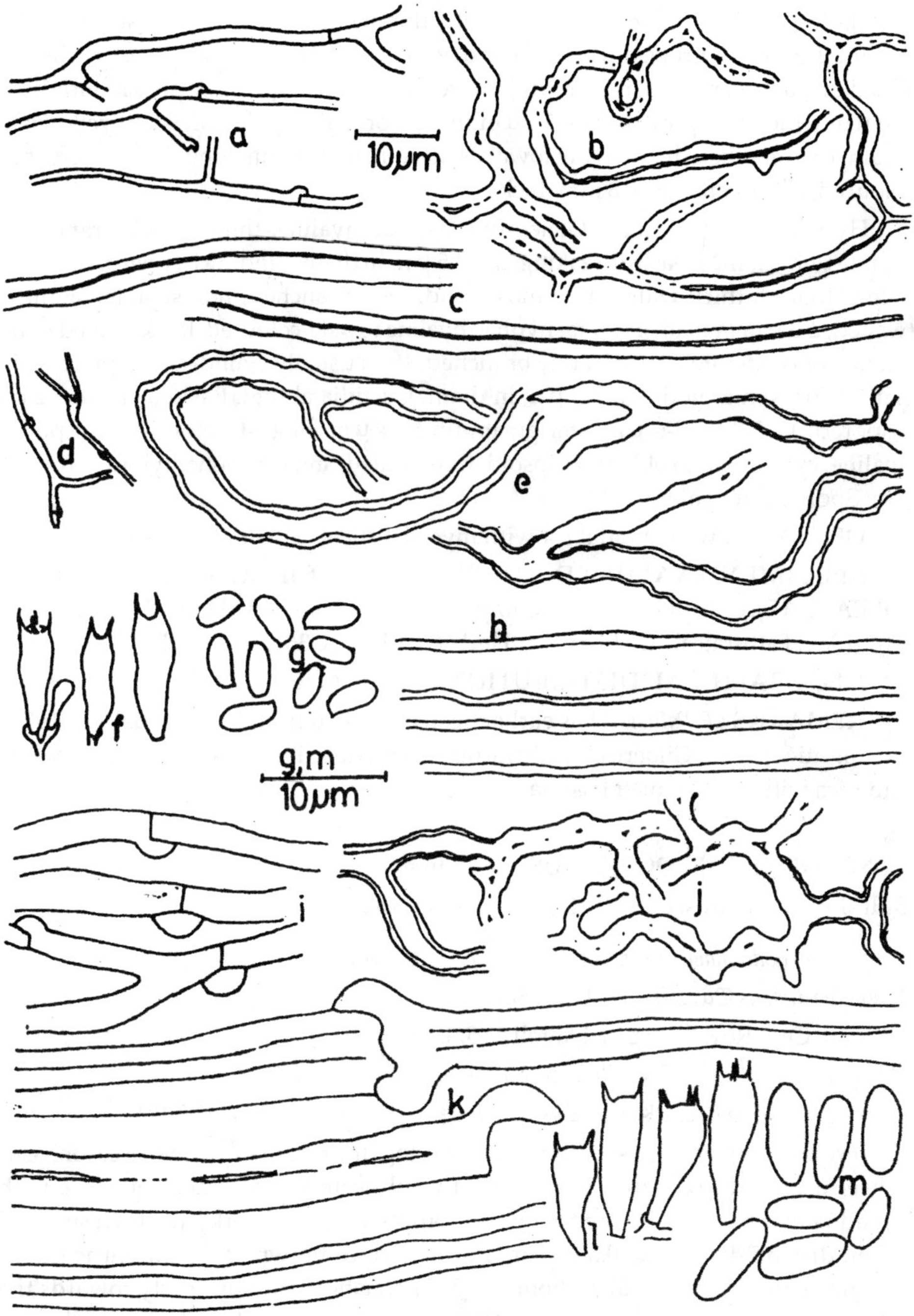

Fig. 17 : a-c *Fomitopsis dochmius* : a-generative hyphae; b-binding hyphae; c-skeletal hyphae; d-h *Fomitopsis feei* : d-generative hyphae; e-binding hyphae; f-basidia; g-basidiospores; h-skeletal hyphae.

i-m *Fomitopsis palustris* : i-generative hyphae; j-binding hyphae; k-skeletal hyphae; l-basidia; m-basidiospores.

undulating towards margin; margin slightly thinner, smooth rigid. Pore surface pale to light grey, more greyish when old or dry, almost even; pores small but visible to naked eye, absent near margin, round to ovoid, regular, 4-5 per mm, pore mouth 150-210 µm wide; dissepiments 60-115 µm thick; pore tubes pale greyish to yellowish cream, darker when dry, narrow towards margin, up to 4 mm long. Context cream-coloured, homogeneous, 1-3 mm thick.

Hyphal system trimitic. Generative hyphae hyaline, thin-walled, branched, septate with clamps, easily breaking, 2.5-3.5 µm in diam. Binding hyphae hyaline to pale-coloured, thick-walled to almost solid, well branched, non-septate, slightly twisted, 2-3 µm in diam. Skeletal hyphae hyaline to pale-coloured, thick-walled with narrow or obliterate lumen, rarely branched, stem usually cylindrical, septate with clamps, broken at nodes, 5-7.5 µm in diam. Cystidia absent. Basidia clavate, 2 or 4-spored, 17-5-20 × 5-6 µm; sterigmata up to 4 µm long, straight. Basidiospores hyaline, cylindrical to oblong-ellipsoid, thin-walled, smooth, nonamyloid, 6-8 × 2-3 µm. Spore print hyaline.

DECAY : Brown cuboidal rot with mycelial sheets in wood crevices.

SPECIMEN EXAMINED : On the trunks of living *Olea dioica* Roxb. (OLEACEAE), Calicut University Campus, Malappuram Dist., Herb. CALI No. G74, 12-7-1984; Patturaical, Trichur Dist., Herb. CALI No. G361, 14-1-1987.

GEOGRAPHICAL DISTRIBUTION : North America, India.

REMARKS : *F. Palustris* has earlier been described from India by Bakshi (1971) as a dimitic species. Gilbertson & Ryvarden (1986) described this species from North America with much longer basidia.

FOMITOPSIS RHODOPHAEUS (Lev.) Imaz.

Bull. Tokyo Sci. Mus. 6 : 92, 1943. Fig. 18 a - c.

Polyporus rhodophaeus Lev., Ann. Sci. Nat. 3, 2 : 190, 1844.

P. cinereo-fuscus Curr., Trans. Linn. Soc. 2, 1 : 124, 1874.

P. aurora Ces., Roy. C. Accad. Napoli 8 : 5, 1879.

P. Semilaccatus Berk., Grevillea 15 : 22, 1886.

Trametes avellana Bres., Krypt. Exsicc. Museo Vindob. Cent 20, 0.157,1910.

Fruitbody pileate, sessile, annual or reviving during the subsequent season, usually effused-reflexed, rarely pileate, mostly imbricate and confluent, attached with broad lateral base, dimidiate to conchate, applanate, hard corky while fresh, rigid when dry; 3.5-10 × 2-6 × 0.5-0.7 cm. Pileus surface brown with concentric butter yellow (4A5), cream (4A3), or Somalis (7E5) zonations, faintly crusty towards the base, otherwise smooth and even, glabrous; margin dull white, becoming brown when bruised, entire, rounded, smooth. Pore surface dull white, with a brownish or greyish shade, surface even, smooth; pores not visible to naked eye, minute, oval to round, regular, 7-8 per mm; pore mouth 85-100 µm wide; dissepiments 25-50(75) µm thick; pore tubes darker than the pore surface, rarely stratified, 2-3 mm long in each layer, sunk into equal depths. Context yellowish white (4A2), homogeneous, 3-4 mm thick fibrous.

Hyphal system trimtic. Generative hyphae hyaline, thin-walled, sparsely branched, septate with clamps, easily breaking at nodes, 2.5-3.5 μm in diam. Binding hyphae hyaline to pale-coloured, branched, non-septate, thick-walled with obliterating lumen or solid, 1.75-3 μm in diam., sparse. Skeletal hyphae hyaline to pale-coloured, unbranched or weakly branched near apices, thick-walled with narrow to broad lumen, cylindrical and straight, slightly flexuous towards extremities, 4-5 μm in diam. Setae absent. Basidia and basidiospores not observed. Large hyaline crystals present in hymenium and dissepiments.

DECAY : Causing brown cuboidal rot.

SPECIMEN EXAMINED : On decaying forest logs of *Cullenia exarilleta* Robyens (BOMBACACEAE), Malakkapara, Sholayar forests, Trichur Dist., Herb. CALI No. G332, 29-8-1986; on dead logs of *Mesua negassarium* (Burm. f.) Kosterm. (CLUSIACEAE), Perumpara (W), Malakkapara, Sholayar forests, Trichur Dist., Herb. CALI No. G348, 30-8-1986; on unidentified forest logs, Kanjarakadavu, Amarambalam Range, Nilambur Forest Division, Malappuram Dist., Herb. CALI No. G104, 16-9-1983; Malakkapara, Sholayar forests, Trichur Dist., Herb. CALI No. G181, 10-6-1985; Mukkali, Silent Valley National Park, Palghat Dist., Herb. CALI No. G229, 15-9-1985.

GEOGRAPHICAL DISTRIBUTION : Mainly an Asian species, known from India, Japan, Philippines, Java, Sri Lanka; one collection from Africa.

REMARKS : Bakshi's (1971) description of *Polyporus rhodophaeus* from West Bengal and Madras agrees well with the present collection except for the smaller pores of the former. Lloyd (1898-1925) had reported this species from India in 1915.

HEXAGONIA Fr.

Fl. Scan. P. 339, 1835.

Fruitbody annual to perennial, pileate, sessile, dimidiate, flabelliform to semicircular, coriaceous, corky or woody. Pileus surface smooth, tomentose or densely hirsute with long dark hairs. Pores entire, angular and mostly large. Context usually thin, dark brown, blackening in KOH. Hyphal system trimitic; generative hyphae hyaline, thin-walled and clamped; binding and skeletal hyphae yellow to golden brown, thick-walled to almost solid, endings of binding and skeletal hyphae often project into the hymenium which then partly becomes a catahymenium; true cystidia absent. Basidiospores cylindrical, longer than 12 μm, smooth, thin-walled and nonamyloid. On deciduous trees, pantropical genus.

Type species : *Hexagonia crinigera* Fr.

Two species have been collected from Kerala.

KEY SPECIES

1. Annual species; thin and coriaceous; pores 10-12 per cm; Causing white rot.. *H. Tenuis*
1′. Perennial species; hard woody; pores 2-3.5 per cm; causing brown rot.. *H. Sulcata*

HEXAGONIA SULCATA Berk.

Lond. J. Bot. 6 : 510, 1847. fig. 18. d-g, Pl., fig.

Daedalea sulcata (Berk.) Ryv., Norw. J. Bot. 24 : 215, 1977.

Fruitbody perennial, mostly solitary, sometimes in groups, pileate, sessile, attaching with a broad lateral base; dimidiate or semi-circular and broadly applanate or slightly irregular and triquetrous, coriaceous and corky while young becoming hard when old, 3-13 × 2-6.5 × 0.8-7 cm. Pileus surface topaz (5C5) to henna (7E8) towards base, fading towards margin, soon becoming blackish; faintly pubescent while young, glabrescent when mature, uneven and rough, usually wrinkled and slightly crusty towards the base, azonate or faintly zonate, somewhat concentrically or irregularly rimose when old; margin smooth, rounded and entire. Pore surface greyish red (8C5) to greyish rose (12B3) while fresh, brownish orange (7C6) to reddish brown (9D5) when dry; pores large, extending to the margin but shallow, angular, 4-6 sided, 2-3.5 per cm; dissepiments up to 1 mm thick; pore tubes stratified or very long, 0.7-5 cm long, inside yellowish white (4A3), concolorous with the pore surface near pore mouth, sunk into almost equal depths, straight, old tubes invaded by context-tissue. Context yellowish white (4A3) to greyish orange (5B4), 1.5-5 mm thick, floccose, homogeneous, of intertwined compact hyphae.

Hyphal system trimitic. Generative hyphae hyaline, slightly thick-walled, branched, septate with clamps, sparsely present in dissepiments, 2-3 µm in diam. Binding hyphae mostly restricted to dissepiments, hyaline, thick-walled with distinct lumen, easily breaking, infrequent, 2.5-3.5 µm in diam. Skeletal hyphae hyaline, thick-walled with narrow to broad lumen, unbranched, non-septate, slightly twisted to flexuous, 3.75-5.5 µm in diam, thick-walled, hyaline, encrusted skeletal hyphae with a slightly swollen apex protruding into the hymenium, 5.5-6 µm thick; encrustations hyaline, more concentrated towards the apex. Basidia and basidiospoores not observed. Hyphal pegs numerous in young sporophores; pale-coloured, pyramidal and acuminate, (30)50-115 × (15)20-3.75 µm projecting beyond the hymenium.

DECAY : Associated with brown cuboidal rot.

SPECIMEN EXAMINED : On decaying forest logs, host unidentified, Panthanthode, Silent Valley National Park, Palghat Dist., Herb. CALI No. G36, 7-9-1983; near abandoned dam site, Silent Valley National Park, Palghat Dist., Herb.

CALI No. G89, 15-9-1985; Poochipara, Silent Valley National Park, Palghat Dist., Herb. CALI No. G363, 15-9-1986.

GEOGRAPHICAL DISTRIBUTION : Sri Lanka, India.

REMARKS : Sporophores of *Hexagonia sulcata* have been collected several times from various localities in North India but this is the first report on its occurrence in South India. The present collection agrees with Bakshi's (1971) description in its macrofeatures and cystidia but the trimitic hyphal system and the presence of hyphal pegs were not reported by him. In the present collections the context is dimitic but the dissepiments are trimitic. The anatomical description of *H. sulcata* from West Bengal (Roy, 1975, 1976) shows similarity with the present materials in its hyphal constitution of context and dissepiments. The description of the type specimen from Sri Lanka (Berkeley, 1847) and a later review of this (Lloyd, 1910a) also agrees with the present collection in its macrocharacters.

HEXAGONIA TENUIS (Hook.) Fr.

Epicrisis p. 498, 1838.

fig. 18. h-l; Pl. fig.

Boletus tenuis Hook., in Kuenth, Syn. Fl. 1 : 10, 1822.

B. reticulatus Hook., Ibid. 1 : 19, 1822.

Hexagonia polygramma Fr., Epicrisis p. 497. 1838.

H. similis Berk., Lond. J. Bot. 5 : 4, 1846.

Hexagonia brevis Berk., Hook. J. Bot. 6 : 229, 1854.

H. muelleri Berk., J. Linn. Soc. Bot. 13 : 166, 1872.

Polyporus changensis Rostr., Bot. Tidsskr. 24 : 360, 1902.

Favolus tenuis (Hook.) Murr., Bull. Torrey Bot. Club 32 : 100, 1905.

Pseudofavolus tenuis (Hook.) G.H. Cunn., N.Z. Dep. Sci. Industr. Res. Bull. 164 : 185, 1965.

Fruitbody annual, sessile, effused-reflexed to pileate, when pileate attached with a broad lateral base, solitary to imbricate, rarely connate; applanate to semicircular or flabelliform, thin and coriaceous, flexible, when fresh and dry; 1.5-6 × 1-3 × 0.2.0.3 cm. Pileus surface bronze (5E5) to shoot brown (5F5) or its allied shades; smooth but uneven, glabrous, concentrically striate to rarely zonate, radially wrinkled, sometimes slightly umbonate at the mid-base; margin entire and thin. Pore suface white to greyish white when fresh; almost topaz (5C5) after handling, greyish beige (4C2) when dry; pores extending to the margin, large, usually hexagonal and regular, or daedaleoid, 10-12 per cm; pore mouth 650-900 μm wide; dissepiments rigid and sharp towards the periphery, 185-250 μm thick; tube insides topaz (5C5) to greyish beige (4C2), 1-1.5 mm long, sunk into equal depths, shallow towards the margin.

Context topaz (5C5) while fresh, light orange (5A4) when dry, homogeneous, up to 1 mm thick.

Hyphal system trimitic. Generative hyphae hyaline, thin-walled, branched, nodose-septate, occasional in context, more frequent in dissepiments, 1.75-2.75 μm in diam. Binding hyphae hyaline, thick-walled, moderately branched, twisted, infrequent in context and dissepiments, 2-2.5 μm in diam. Skeletal hyphae hyaline, thick-walled with lumen obliterating or narrow, unbranched and non-septate, usually cylindrical and straight but flexuous and slightly twisted towards extremities, dominating, 2.5-3.75 μm in diam. Basidioles numerous, hyaline, cylindro-clavate, thin-walled, 20-27.5 × 6.25-7.5 μm. Cystidia none. Basidia not observed. Basidiospores hyaline, cylindrical to sub-allantoid, thin-walled, smooth, 0 to 2-guttulate, nonamyloid, 9-14 (15.5) × 3-4.5 μm. Spore-print white.

DECAY : White fibrous rot.

SPECIMEN EXAMINED : On dead branches of *Anacardium occidentale* Linn. (ANACARDIACEAE), Botanical Garden, Calicut University Campus, Malappuram Dist., Herb. CALI No. G88, 6-7-1984, G90, 30-2-1984; on fallen branches of *Calycopteris floribunda* (Roxb.) Poir. (COMBRETACEAE), near Calicut University Campus, Malapuram Dist., Herb. CALI No. G327, 11-8-1985; on dead branches of *Moringa oleifera* Lam. (MORINGACEAE), Calicut University Campus, Malappuram Dist., Herb. CALI No. G382, 20-11-1987; on fallen stumps of *Mangifera indica* Linn. (ANACARDIACEAE), Pazhayangadi, Cannanore Dist., Herb. CALI No. G393, 31-12-1987; on forest logs, host unidentified, Karadimala, Brahmagiri Ranges, Tirunelli, Wyanaad dist., Herb. CALI No. G624, 2-10-1985.

GEOGRAPHICAL DISTRIBUTION : Central and South America, Mexico, Yucatan, West Indies, Africa, India, Sri Lanka, Philippines, Fiji, Australia, New Zealand.

REMARKS : Great variation in size of spores of *H. tenuis* collected from various parts of the world has been reported by Parmasto (1983a). The shape and size of the spores as well as macrocharacters of the present collection agrees well with those in Bakshi's (1971) description. This polypore is more common in inhabited areas while only one collection was from an undisturbed evergreen forest.

IRPEX Fr.

Elench. Fung. 1: 142, 1828.

Fruitbody annual, sessile, effused-reflexed or resupinate. Pileus surface white to palebuff, tomentose to hirsute. Pore surface strongly hydnaceous. Hyphal system dimitic; generative hyphae simple-septate. Cystidia thick-walled, encrusted. Basidiospores cylindric, negative in Melzer's. Decay a white rot of dead hard wood and conifers. Monotypic genus. Type species : *Irpex lacteus* (Fr. ex Fr.) Fr.

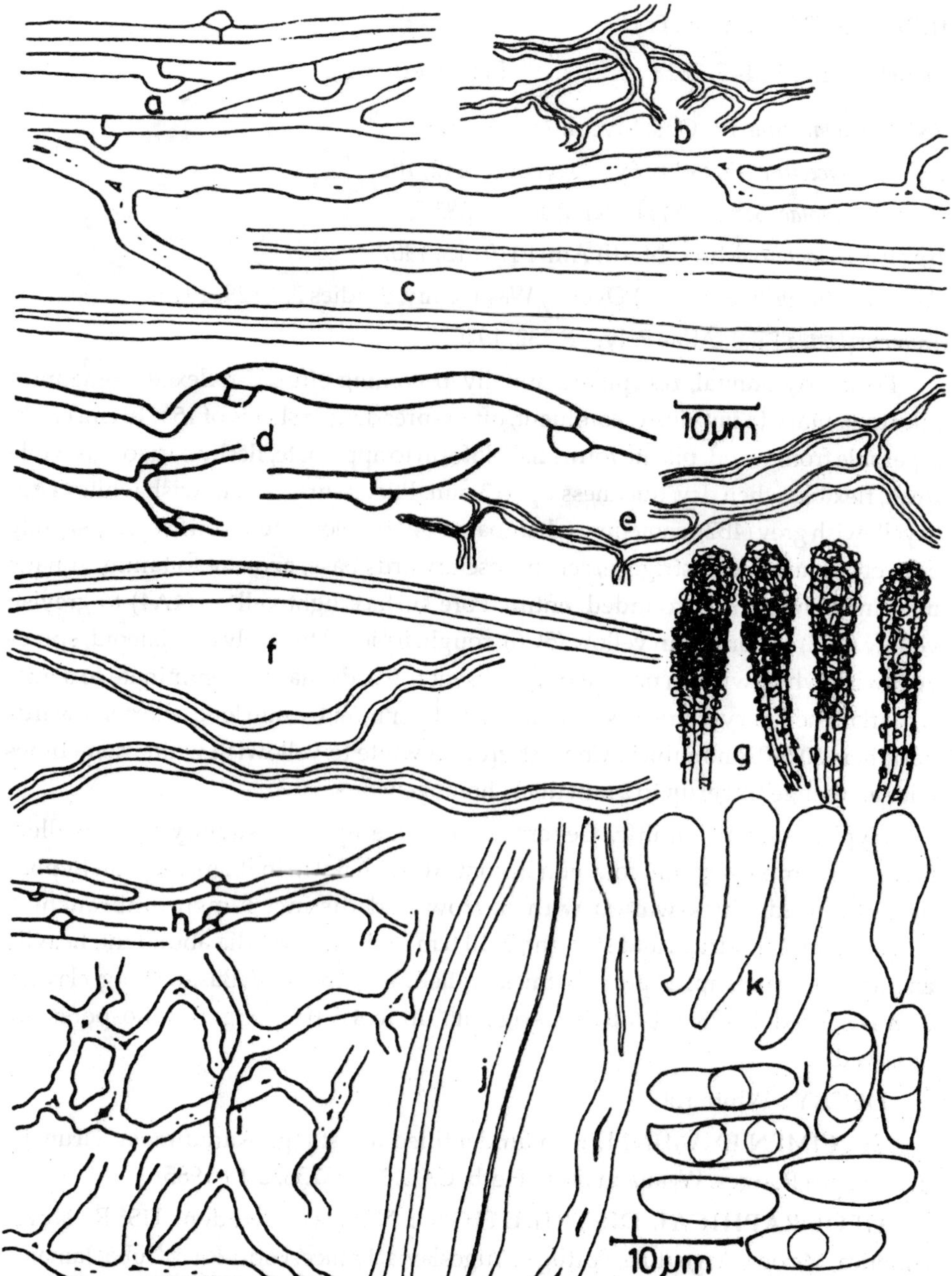

Fig. 18 : a-c *Fomitopsis rhodophaeus* : a-generative hyphae; b-binding hyphae; c-skeletal hyphae;

d-g *Hegagonia sulcata* : d-generative hyphae; e-binding hyphae; f-skeletal hyphae; g-encrusted skeletal hyphae at pore mouth.

h-l *Hexagonia tenuis* : h-generative hyphae; i-binding hyphae; j-skeletal hyphae; k-basidioles; l-basidiospores.

IRPEX LACTEUS (Fr. ex Fr.) Fr

Elench. Fung. 1 : 145, 1828. Fig. 19a-d., fig

Sistotrema lacteum Fr., Obs. Mycol. 2 : 266, 1818

Hydnum lacteum (Fr. ex Fr., Syst. Mycol. 1 : 412, 1821.

Irpex bresadolae Schul., Ms II, Nr. P. 1348, 1885.

Irpiciporus lacteus Murr., North Am. F1. 9 : 15, 1907.

Polyporus tulipiferae (Schw.) Overh., Wash. Univ. Studies 3, 1 : 29, 1915.

Trametes lactea Pilat, Atlas Polyp. P. 332, 1940.

Fruitbody annual, resupinate initially, becoming effused-reflexed, sometimes pileate, solitary to imbricate, confluent, often spreading in sheets of 15 × 10 cm. easily separable from wood, pileate forms flabelliform to applanate, fleshy coriaceous while fresh, flexible when dry, thickness up to 3 mm. Pileus surface yellowish white (4A2) to yellowish grey (4B2), sometimes champagne (4B4), colour fading on keeping; slightly uneven, velutinate to strigose, scrupulose towards base, azonate to faintly zonate; margin slightly thick, rounded, entire. Pore surface light yellow (4A4) to greyish yellow (4B2) to Pompeian yellow (5C6); rough, irpicoid to rarely daedaleoid, spines yellowish white with an orange tinge, absent towards margin, acuminate to somewhat flattened or cylindrical, scattered, up to 1 per mm; 1-2 mm long, shorter towards periphery, 0.5-0.7 mm thick. Context greyish white to yellowish white, sometimes with an orange tinge; up to 2 mm thick, homogeneous.

Hyphal system dimitic. Generative hyphae hyaline, slightly thick-walled, branched, simple-septate, common in context, 3.5-6.5 μm in diam. Skeletal hyphae pale-coloured, thick-walled with narrow and distinct lumen, unbranched, non-septate, cylindrical and straight, 5-10 μm in diam. Cystidia abundant, heavily encrusted towards apical portion, thick-walled, 4-8 μm broad. Basidia long, clavate, 2 or 4- spored, 21-26 X 4-6 μm; sterigmata up to 4.5 μm long. Basidiospores not observed.

DECAY : White rot.

SPECIMEN EXAMINED : On unidentified forest logs, Karadimala, Tirunelli, Brahmagiri Ranges, Wyanaad Dist., Herb. CALI No. G256, 2-10-1985.

GEOGRAPHICAL DISTRIBUTION : Norway, Sweden, USSR, North America, Mexico, Argentina, Portugal, Yugoslavia, France, Porto-Rico, Cuba, Jamaica, India.

REMARKS : The irpicoid hymenophore with encrusted cystidia and simple-septate hyphae are most diagnostic of this species. In the earlier Indian report of *I. Lacteus*, Bakshi (1971) describes it as having a poroid hymenophore with hyphal pegs.

JUNGHUHNIA Corda emend Ryv.

Anl. Stud. Mycol. P. 195, 1842; Persoonia 7 : 18, 1972.

Fruitbody annual, resupinate, rarely effused-reflexed. Pore surface cream-coloured to pinkish buff or cinnamon; pores mostly small, with thin, lacerate dissepiments. Hyphal system dimitic; generative hyphae with clamps; thick-walled skeletocystidia rare to abundant, heavily, encrusted, embeded or projecting. Basidia clavate, 4 sterigmate. Basidiospores ovoid to cylindric, curved in some species. Causing white rot of dead hardwood and conifers.

Type species : *Junghuhnia crustacea* (Jungh.) Ryv.

Only one species has been recorded from Kerala.

JUNGHUHNIA CRUSTACEA (Jungh.) Ryv.

Persoonia 7 : 18, 1972. fig. 19. e-i; Pl., fig.

Laschia crustacea Jungh., Verh. Batav. Genootsch. 17 : 75, 1838.

Poria crustacea (Jungh.) Bres., Ann. Mycol. 8 : 587, 1910.

Grammothele delicata Bres., Hedwigia 56 : 299, 1915.

Fruitbody annual, resupinate, up to 1 mm thick, easily separable from wood, margin undifferentiated, reflexed about a few mm or adpressed, spongy while fresh, semi-brittle when dry; spreading in sheets of about 20 × 20 cm, occurring often in small isolated patches. Pore surface white while fresh, yellowish white (4A2) when dry, slightly rough and uneven; pores small but visible to unaided eye, absent near margin, scattered, angular, 4-6 per mm; pore mouth 125-150 µm wide; dissepiments 75-100(150) µm thick, dentate; pore tubes shallow, white, less than 0.5 mm long. Context white, 200-250 µm thick.

Hyphal system dimitic. Generative hyphae hyaline, thin-walled, branched, septate with clamps, common in context, 2.5-4 µm in diam. Skeletal hyphae hyaline, to pale coloured, thick-walled, mostly unbranched, cylindrical and equal in width, 3-4.5 µm in diam. Basidia long, clavate, 2 or 4-spored, 15-20 × 4.5-5.5 µm; sterigmata to 3 µm long. Skeletocystidia numerous; long, cylindrical, encrusted at the apex, encrustations dissolving in KOH, clavate, 25-30 × 10-12.5 µm. Basidiospores hyaline, globose to sub-globose, thin-walled, smooth, eguttulate, nonamyloid, 4-4.5(5) × 3(3.5) µm.

DECAY : White rot.

SPECIMEN EXAMINED : On decomposed forest logs, host not identified, Chembra Peak, Meppadi, Wyanaad Dist., Herb. CALI No. G234, 11-9-1985.

GEOGRAPHICAL DISTRIBUTION : Philippines, Malaysia, Krakatau, Long Island, Tanzania, India.

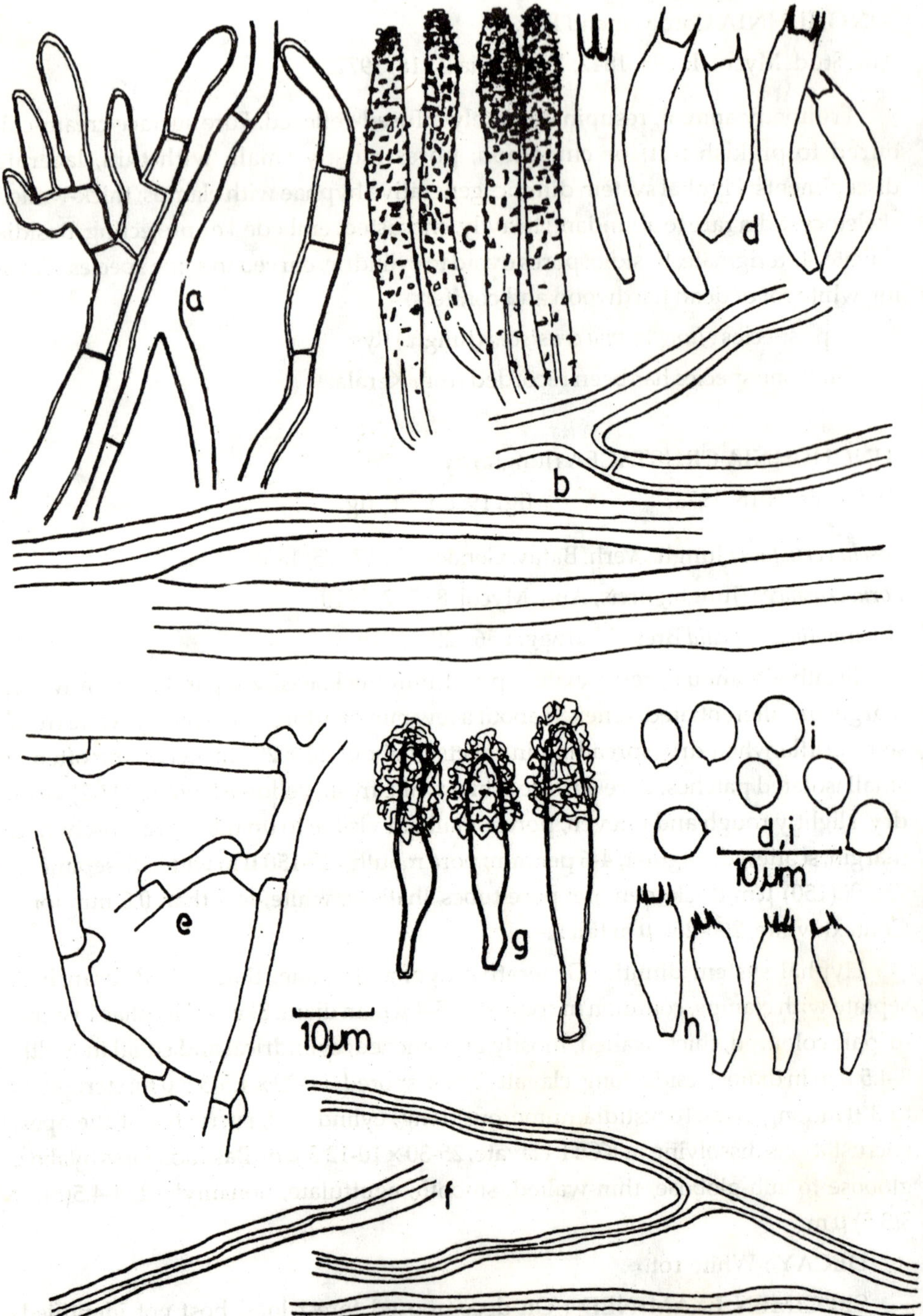

Fig. 19 : a-d *Irpex lacteus* : a-generative hyphae; b-skeletal hyphae; c-cystidia; d-basidia. e-i *Junghuhnia crustacea* : e-generative hyphae; f-skeletal hyphae; g-skeletocystidia; h-basidia; i-basidiospores.

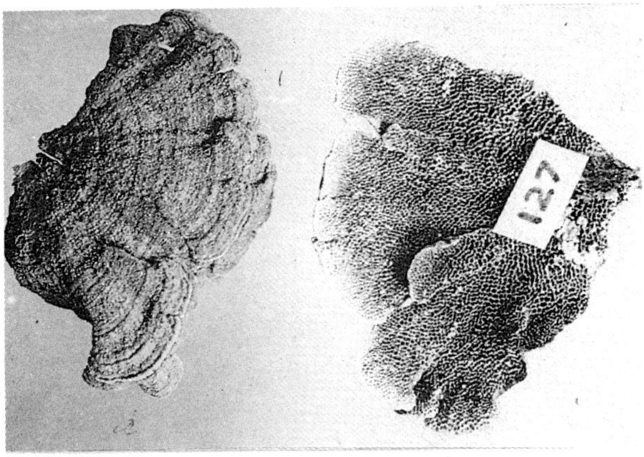

1. Cyclomyces setiporus—(a) Pileus (b) pore surface

2. Cyclomyces tabacinus—pileus surface

3. Phellinus rinosus—pileus surface

4. Phellinus troyanus—pileus surface

5. Phellinus contiguis—pore surface

6. Phellinus fastuosus—pileus surface

7. Phellinus ferreus—pore surface

8. Phellinus gilvoides—pore surface

9. Phellinus hohnelii—pileus surface

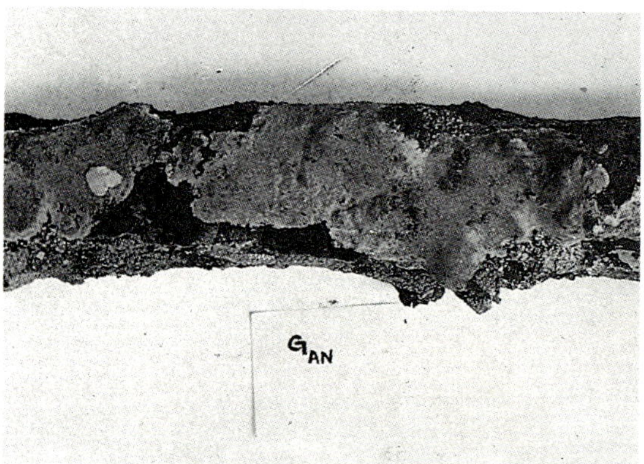

10. Phellinus punctatus—pore surface

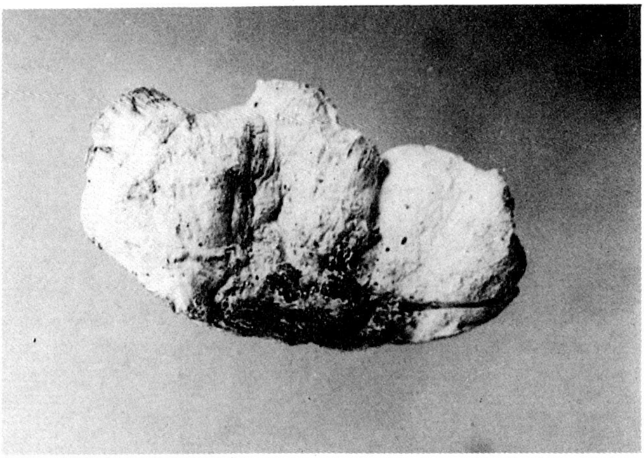

11. Bjerkandera sp.—pileus surface

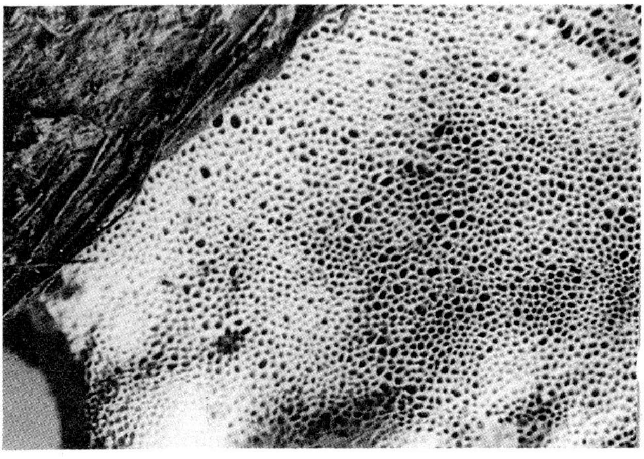

12. Bjerkandera sp.—pore surface

13. Coriolopsis sanguinaria—habit

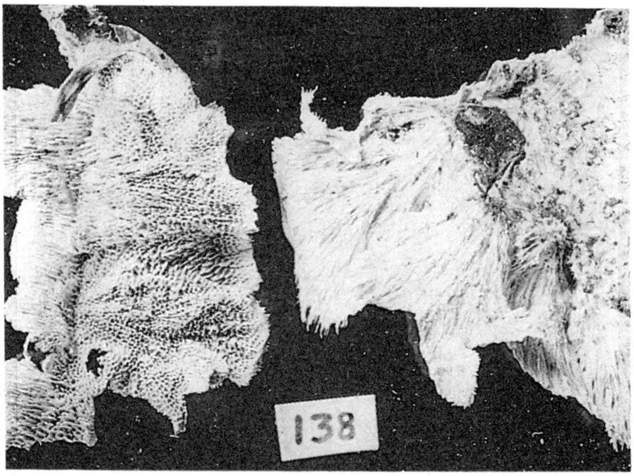

14. Coriolopsis telfari—pileus and pore surface

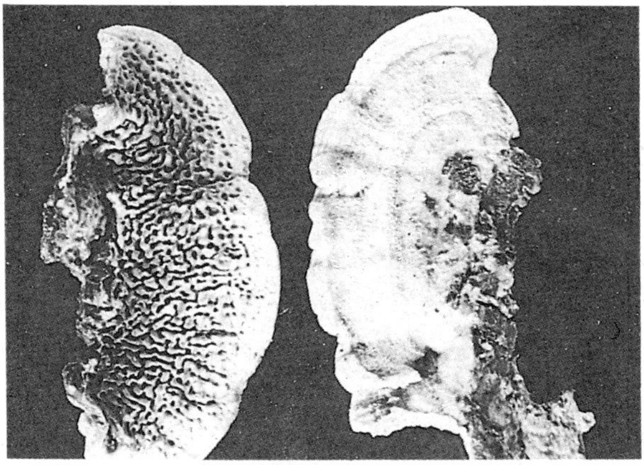

15. Flavadon flavus—pileus and pore surface

16. Fomitopsis dochmius—pileus surface

17. Hexagonia sulcata—pileus surface

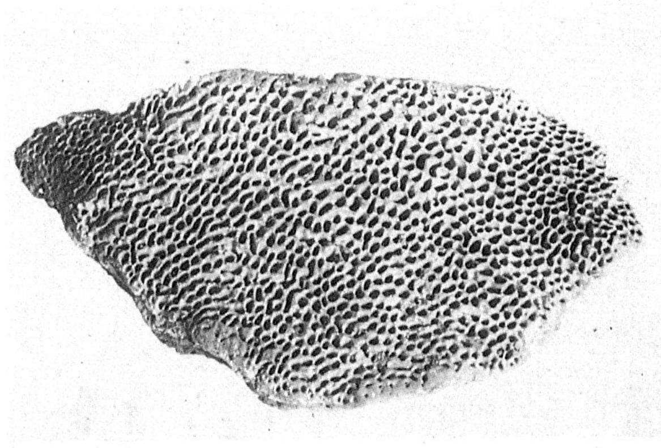

18. Hexagonia sulcata—pore surface

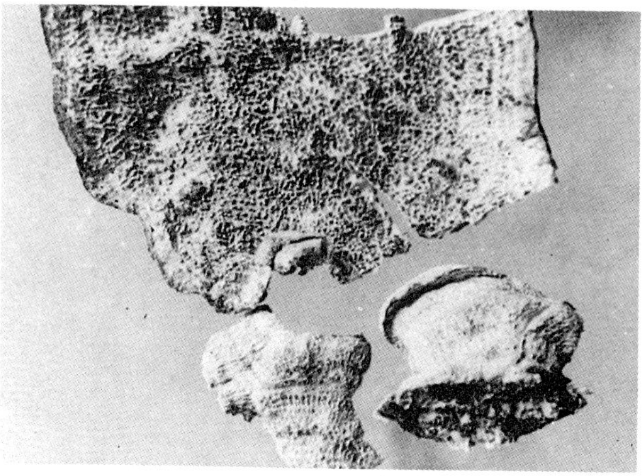

19. Irpex acteus—pileus and pore surface

20. Lenzites sp.—pileus surface

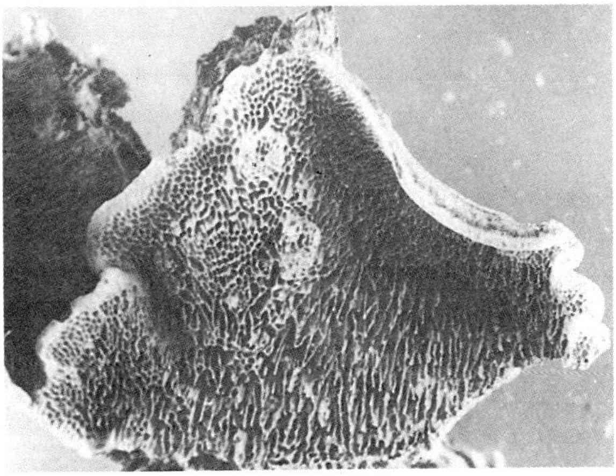

21. Lenzites sp.—pore surface

22. Nigroporus niger—pore surface

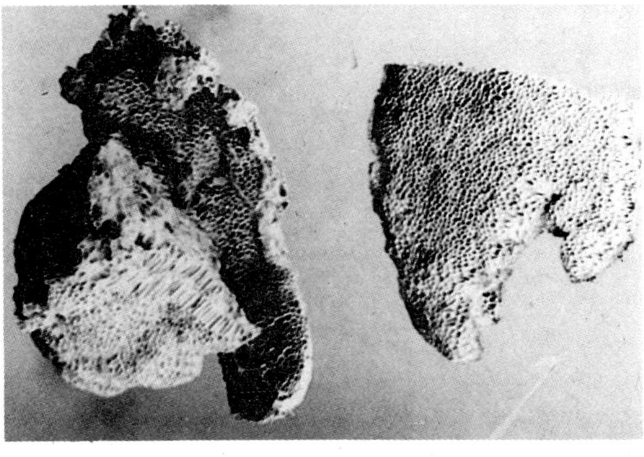

23. Oxyporus mollissimus—pileus and pore surface

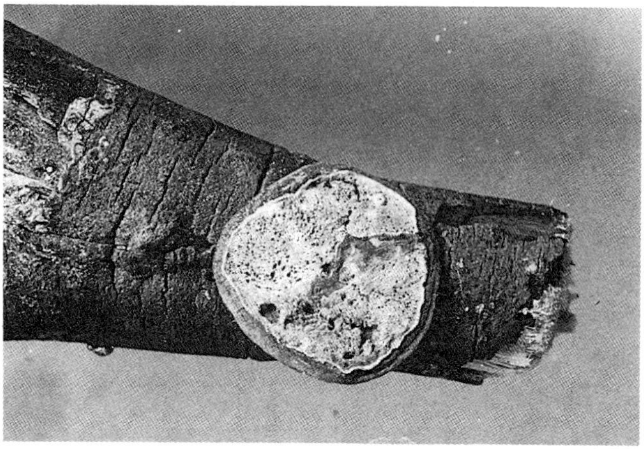

24. Piloporia indica—Habit

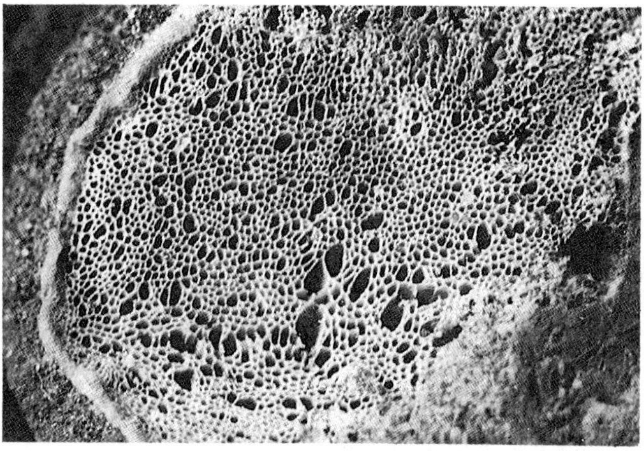

25. Piloporia indica—pore surface

26. Polyporus grammocephalus—Habit

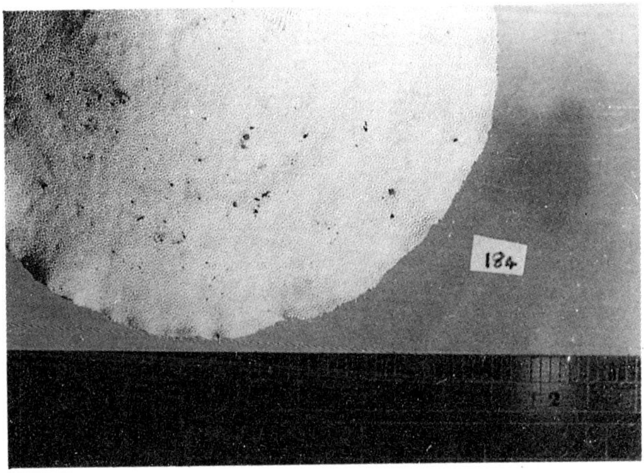

27. Polyporus grammocephalus—pore surface

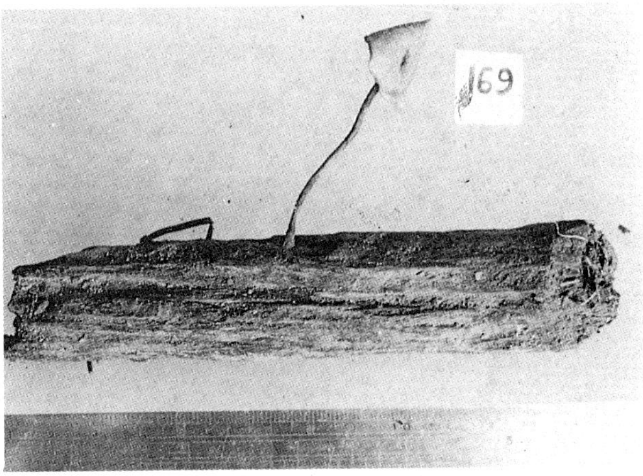

28. Polyporus hemicapnodes—Habit

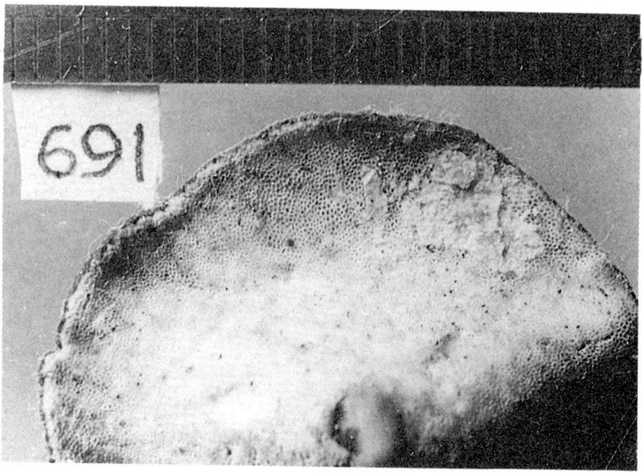

29. Polyporus hemicapnodes—pore surface

30. Rigidoporus microporus—pileus surface

31. Rigidoporus microporus—pore surface

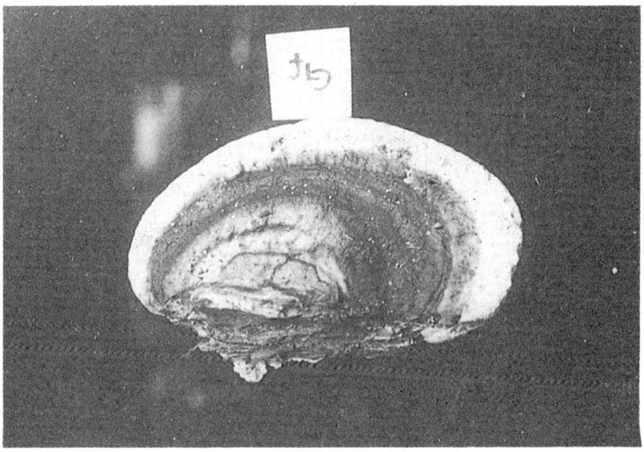

32. Trametes cingulata—pileus surface

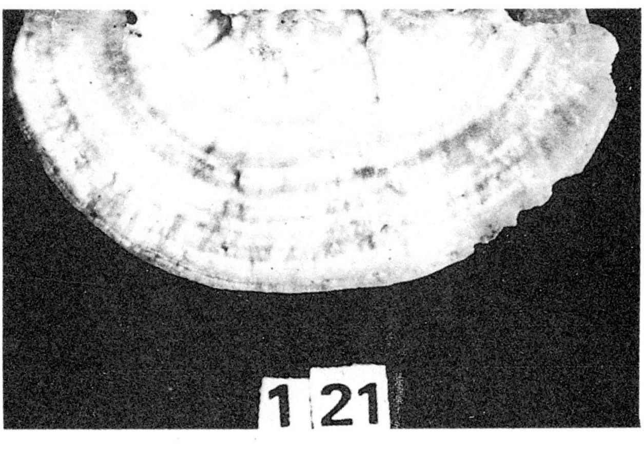

33. Trametes cotonea—pileus surface

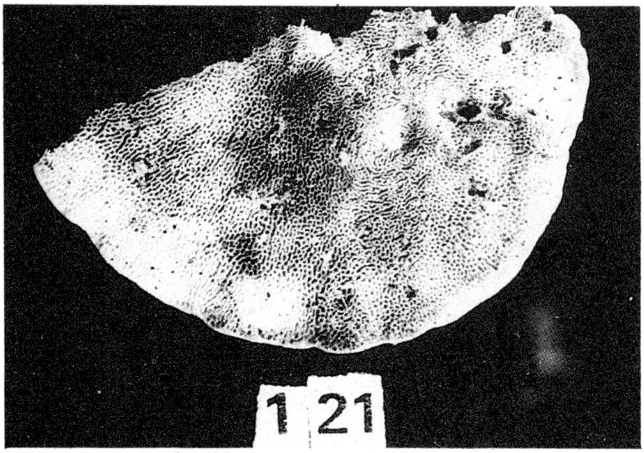

34. Trametes cotonea—pore surface

35. Trametes marianna—Habit

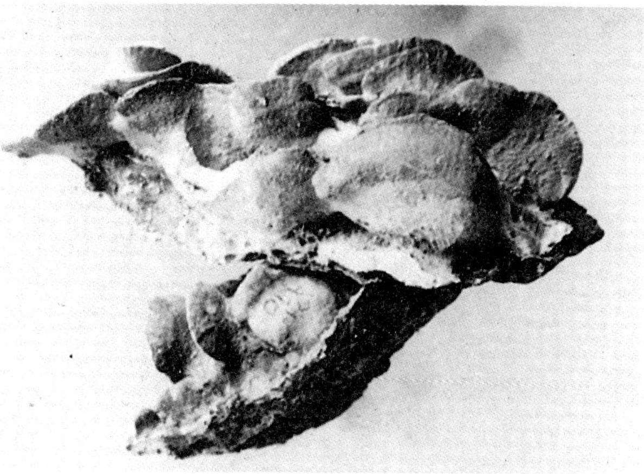

36. Trametes modesta—Habit

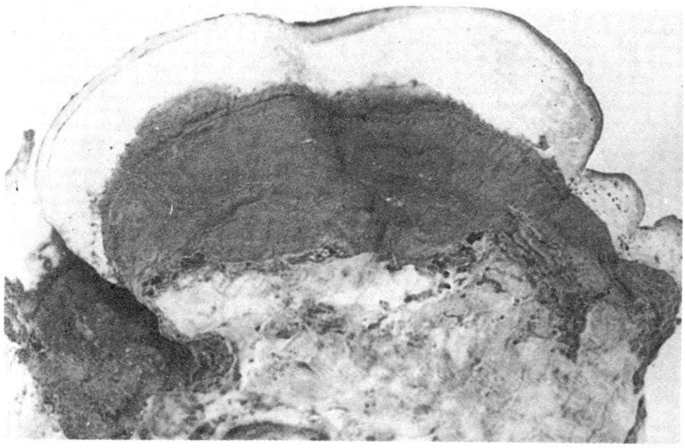

37. Trametes scabrosa—pileus surface

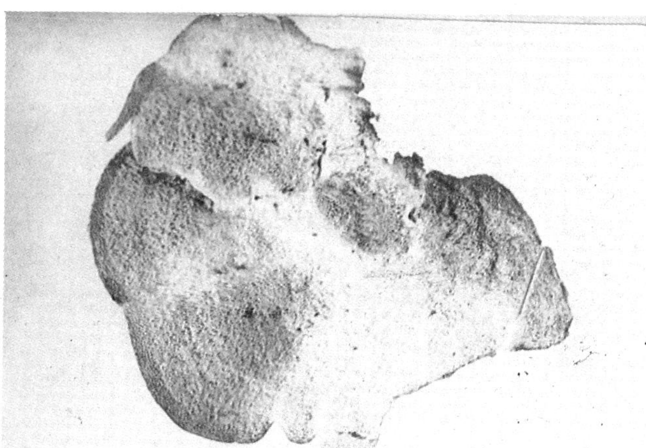

38. Trametes scaborsa—pore surface

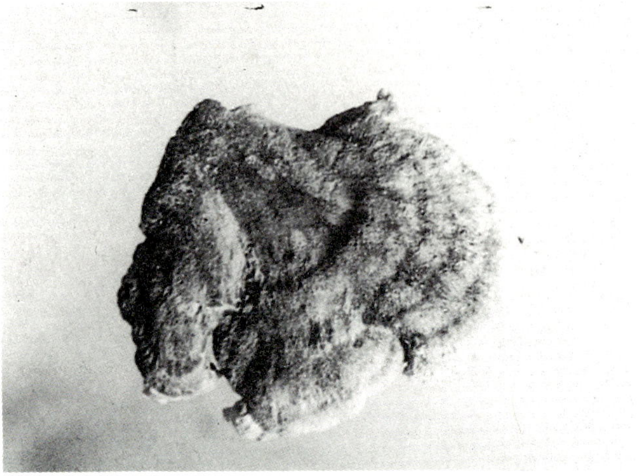

39. Tyromyces pelliculous—pileus surface

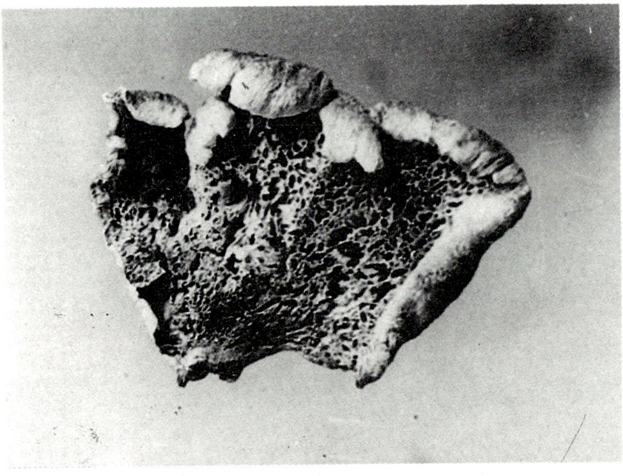

40. Tyromyces pelliculous—pore surface

REMARKS : This is the first report on the occurrence of *J. crustacea* in India. The present collection agrees closely with the description of this polypore from North America by Lowe (1963). Ryvarden & Johansen (1980) reported oblong-ellipsoid spores for their East African collection of *J. crustacea.*

LENZITES Fr.

Fl. Scand. P. 339, 1835.

Fruitbody annual, pileate. Pileus surface white to greyish with age, glabrous to hirsute, often in distinct zones. Pore surface white to yellowish, lamellate to daedaleoid. Context concolorous with the pore surface, tough and moderately thick. Hyphal system trimitic; generative hyphae hyaline, thin-walled, clamped; binding hyphae hyaline, strongly branched and tortuous, in the trama below the hymenium they have long, sword-like side branches more or less parallel and pointing towards, and in parts into, the hymenium, but never above it; skeletal hyphae hyaline, thick-walled, to solid; basidiospores hyaline, cylindrical, smooth, thin-walled, nonamyloid. On deciduous wood causing white rot, cosmopolitan genus.

Type species :*Lenzites betulina* (L. ex. Fr.) Fr.

Three species have been recorded from Kerala.

KEY TO SPECIES

1. Cystidia absent.......... *L. Acuta*
1′. Cystidia present.......... 2
2. Binding and skeletal hyphae protruding sword-like into the hymenium but not encrusted.......... *L. Betulina*
2′. Binding and skeletal hyphae encrusted if protruding into the hymenium or encrusted cystidia present.......... *L. Taxonomic* species.

LENZITES ACUTA Berk.

Hook. Lond. J. Bot. 1 : 146, 1842.

fig. 20. l-d; Pl. , fig.

Daedalea tenuis Berk, Hook. Lond. J. Bot. 1 : 151, 1842.

Polyporus poecilus Berk., Ann. Nat. Hist. 10 : 372, 1843.

P. laeticolour Berk. Ibid. 10 : 374, 1843.

Daedalea inequabilis Berk. Ibid. 10 : 374, 1843.

D. lurida Lev., Ann. Sci. Nat. 3, 2 : 198, 1844.

D. flavida Lev., Ibid. 3, 2,. 198, 1844.

D. pruinosa Lev., Ibid. 3, 2 : 198, 1844.

Hexagonia glabra Lev., Ibid. 5 : 143, 1846.

Daedalea macrozona Lev., Ibid. 5 : 142, 1846.

Trametes colliculosus Berk., Hook. Lond. J. Bot. 6 : 506, 1847.

T. lobata Berk., Hook. J. Bot. 3 : 84, 1851.

Lenzites ochrophylla Berk., Ibid. 6 : 133, 1854.

L. exima Berk. & Curt, Ibid. 6 : 134, 1854.

Polyporus lenziteus Lev., in Zollinger, Syst. Verzeichnis p. 17, 1854.

Lenzites becklerii Berk., J. Linn. Soc. Bot. 13 : 161, 1872.

Daedalea hobsonii Berk., Ibid. 13 : 165, 1872.

D. aulacophyllus Berk. & Br., Ibid. 13 : 166, 1872.

D. subsulcata Berk. & Br., Ibid. 14 : 56, 1873.

D. glabrescens Berk., Ibid. 16 : 39, 1878.

Lenzites alutacea Cke., Grevillea 11 : 121, 1882.

Daedalea scalaris Berk. & Br., Trans. Linn. Soc. Bot. 2, 2 : 1883.

Lenzites malaccensis Sacc. & Cub., Syll. Fung. 5 : 645, 1887, nom. Nov. Pro *L. Platyphylla* Cke., non Lev.

Daedalea schomburghii Berk. ex Sacc., Syll. Fung. 6 : 376, 1888, based on the specimens as *D. hobsonii* (cf. Lloyd, 1910).

D. flabellum Berk. ex Sacc., Ibid. 9 : 200, 1891.

D. suberosa Mass., Bull. Misc. Inf. Kew 1906 p. 94, 1906. Nom. Nov. Pro *D. Ochracea* Mass., non Kalch.

D. isabelline Murr., Bull. Torrey Bot. Club 35 : 414, 1908.

D. subconfragosa Murr., Ibid. 35 : 415, 1908.

Lenzites submurina Murr., Ibid. 35 : 416, 1908.

L. adusta Mass., Bull. Misc. Inf. Kew 1910 p. 250, 1910.

Trametes transmutans Lloyd, Mycol. Writ. 6 : 996, 1920.

Hexagonia flavofusca Lloyd, Ibid. 7 : 1156, 1922.

Fruitbody epixylpous, annual often reviving during the season, pileate and sessile, attached with a broad lateral base, solitary to imbricate, rarely confluent, dimidiate, semi-circular, flabelliform or irregular, applanate, corky while fresh, tough when dry; 3.5-30 × 2.2-20 × 0.5-3 cm. Pileus surface light yellow (4A4) or maize (4A6) or dark blonde (5D6), concentrically striate, finely velutinate while young, becoming glabrescent with age; margin smooth, regular. Pore surface yellowish white (4A2) to light yellow (4A4) to golden brown (5D7), poroid to daedaleoid to lamelloid, intermixed types frequent, sometimes margin distinct, otherwise dissepiments extending to the margin; pores in radial rows, shape varies form round to rectangular or daedaleoid, up to 5 per cm; dissepiments (lamellae) 0. 5- 1 . 5 mm thick, concolorous with the pore surface, edges often sharp, 0.4-2.5 cm long. Context pastel yellow (3A4) to almost concolorous with the pore surface, 1. 5 mm thick, homogeneous.

Hyphal system trimitic. Generative hyphae hyaline, thin-walled, branched, septate with clamps, 2.25-3 µm in diam. Binding hyphae pale-coloured, thick-walled with distinct lumen or often solid, highly branched, non-septate, 2. 5-4 µm in diam. Skeletal hyphae pale-coloured, thick-walled, unbranched, cylindrical and straight, extremities slightly attenuated, non-septate, 4. 5-6.5 µm in diam. Cystidia none. Basidia not observed. Basidiospores hyaline, broadly cylindrical, thin-walled, smooth, eguttulate, nonamyloid. 7-7.5 × 3-3.75 µm; spore print white.

DECAY : White spongy rot, on decorticated wood or dead branches of living trees.

MATERIALS EXAMINED : On fallen branches of *Careya arborea* Roxb. (LECYTHIDIACEAE), Pullamala, Brahmagiri Ranges, Tirunelli, Wyanaad Dist., Herb. CALI No. G260, 2-10-1985; on decaying logs of *Artrocarpus heterophyllus* Lam., (MORACEAE), Pazhayangadi, Cannanore Dist., Herb CALI No. G319, 31-12-1987; on decaying *Bambusa sp.* stumps (POACEAE), Kerala Forest Research Institute Sub-Centre, Nilambur, Malappuram Dist., Herb. CALI No. G39, 16-9-1983; on deacying *Tectona grandis* Linn. logs (VERBENACEAE), Botanical Garden, Calicut University Campus, Malappuram Dist., Herb. CALI No. G109, 5-8-1983; on unidentified forest logs, Sholayar forest, Trichur Dist., Herb. CALI No. G9, 28-6-1983; Panthenthode, Silent Valley National Park, Plaghat dist., Herb. CALI No. G35, 7-9-1983; Kaduvakkunnu, Nadugani forests, Nilambur Forest Division, Malappuram Dist. Herb CALI No. G44, 28-1-1984, G132, 13-11-1984; Vazhachal, Sholar forests, Trichur Dist., Herb. CALI No. G82, 26-7-1984; Kanjirakadavu, Nilambur forests, Malappuram Dist., Herb., Herb. CALI No. G110. 30-6-1983; Chooralmala, Meppadi, Wyanaad Dist., Herb. CALI No. G252, 13-9-1985; on decaying wood, Calicut University Campus, Malappuram Dist., Herb. CALI No. G98, G99, 11-6-1984; Nilambur forests, Herb. FRI No. 4287, K. Bagchee, 29-4-1946; Cannanore Dist., Herb. FRI No. 4348, K. Bagchee, 3-6-1946; Wyanaad Dist., Herb. FRI No. 5146, K. Bagchee, 26-4-1950.

GEOGRAPHICAL DISTRIBUTION : East Africa, New Guinea, New Zealand, Australia, Philippines, China, Pakisthan, India.

REMAKRS : This fungus was earlier known from India as *Daedalea flavida* (Roy, 1968b; Bakshi, 1971). A recent revision of polypores described by Berkeley and Leveille by Ryvarden (1976c, 1981) lists a series of synonyms of this common polypore and allocates priority to *Lenzites acuta* Berk. Lloyd (1898-1925) reported this fungus as the 'most frequent species in the East' and according to him 'every collection has different hymenial form'. Ryvarden (1976c) described this as a species 'which has been repeatedly described as new because of its variable hymenophore which can be poroid, daedaleoid, labyrinthine or irpicoid to partly lamellate, often within the same specimen'. Some of the East African collections (Ryvarden & Johansen, 1980) have been described as almost stipitate with smaller pores.

The polymorphic fruitbodies of *L. Acuta* are common in Kerala in all types of habitats, including forest, wooded areas, gardens, timber depots and exposed wooden structures of buildings.

LENZITES BETULINA (L. ex Fr.) Fr.
Epicrisis p. 405, 1838.

fig. 20. e-h; Pl. fig.

Agaricus betulimus L., Sp. Plant. P. 1176, 1783.)
Daedalea betulina (L.) ex. Fr., Syst. Mycol. 1 : 333, 1821.
Lenzites ochraceus Lloyd, Mycol. Notes 66 : 1130, 1922.
L. variegata Lloyd, Ibid. 72 : 1273, 1924.

Fruitbody annual, pileate, sessile, attached with a narrow converging base, solitary to imbricate dimidiate to flabelliform, applanate, coriaceous while fresh, tough when dry; 3.5-6.5 × 2.5-4 × 0.3-0.5 cm. Pileus surface yellowish white (4A2) to putty

(AB2) while fresh, orange grey (5B2) to more brownish when dry; finely tomentose to hirsute, concentrically zonate and ridged, zonations often close; margin thin, entire and smooth. Pore surface yellowish white (4A2) to creamy white (4A3); mostly lamellate, often daedaleoid in some areas, lamellae regular, length varying, sometimes forked near the margin 1-1.5 per mm, 1.5-3 mm long, up to 450 μm thick at the base; edges thin, entire. Contest white to creamy white (4A3), fibrous, homogeneous, 1-2mm thick.

Hyphal system trimitic. Generative hyphae hyaline, thin-walled, branched, septate with clamps, 2-2.5(4) μm in diam., sparsely present in context. Skeletal hyphae hyaline to pale-coloured, thick-walled with broad, narrow or obliterate lumen, unbranched, non-septate or rarely simple-septate, 3.5-7.5 μm in diam. Binding hyphae hylaine to pale-coloured, thick-walled with narrow and distinct lumen, branched, (3) 5-7.5 μm in diam. Binding hyphae in dissepiments more closely branched and projecting into the hymenium with sharp apices, appearing like cystidia, 5-7.5 μm in diam. Basidia and basidiospores not observed.

DECAY : Associated with white rot.

SPECIMEN EXAMINED : On decaying forest logs, host not identified, Vellarimala, Chembra Peak, Meppadi, Wyanaad Dist., Herb. CALI No. G125, 23-9-1984.

GEOGRAPHICAL DISTRIBUTION : Consmopolitan, but common in the temperate boreal zone than in the tropical zone.

REMARKS : Bakshi (1971) has reported the occurrence of this species on a large number of hardwoods and conifers from the Himalayan region while only a single collection was made from Kerala in the course of this survey.

LENZITES taxonomic species.

fig. 20. l-n; fig.

Fruitbody annual, pileate, sessile, attached laterally with a broad base, sometimes effused-reflexed, imbricate and confluent; usually semi-circular, applanate, coriaceous when fresh, more rigid when dry; 1-5-3.5 × 1-2 × 0.2-0.4 cm. Pileus surface straw yellow (3B4) to blonde (4C4), sometimes with greyish orange (5B4) zonations, white or greyish orange (5B4) to brownish grey (5F8) towards margin; finely tomentose with glabrous concentric zones, hairs agglutinated near the base, margin entire, smooth and rounded. Pore surface caramel (brown) (6C6) to sahara (6C5), sometimes orange white (6A2) towards margin; daedaleoid to sinuate or irpicoid, sometimes poroid towards the periphery, margin sterile; pore or spines 1-1.5 per mm; dissepiments 200-250 μm thick; spines usually flattened, concolorous with the pore surface, 1.5-2 mm long, sunk into almost unequal depths. Context sepia (5F4) to dark brown, homogeneous, fibrous, 1-1.5 mm thick.

Hyphal system trimitic. Generative hyphae hyaline, thin to slightly thick-walled with broad lumen, usually simple-septate, clamps rarely present in dissepiments, 2-4 μm in diam. Binding hyphae pale brownish yellow, thick-walled with definite lumen, nonseptate, well branched, rarely present in all tissues, 2.5-3.5 μm in diam. Skeletal hyphae pale brownish yellow, thick-walled with broad lumen, mostly unbranched, non-septate, rare in dissepiments, 4.5-6.5 μm in diam. Skeletal and binding hyphae with apical encrustration often protruding into the hymenium or cystidiate, cystidia subulate to obclavate with apical encrustations, thick-walled, 32-41 × 8-12 μm. Basidia clavate, 2 or 4-spored, 17-20 × 6-6.5 μm; sterigmata 2-3 μm long. Basidiospores hyaline, short cylindric to sub-ellipsoid, sometimes obovate; smooth, nonamyloid, eguttulate, 6.5-8 × 3.5-4 μm.

DECAY : White rot.

MATERIALS EXAMINED : On decaying logs of *Artocarpus heterophyllus* Lam., from a wooded area, Pazhayangadi, Cannanore Dist., Herb. CALI No. G384, 10-12-1987, G394, 31-12-1987.

REMARKS : The present collection is close to *Lenzites steroids* (Fr.) Ryv. but differs in the colour of the pileus and pore surfaces, type of cystidia and spore characters. Though it resembles *Daedalea* pers. ex. Fr. in some respects, it differs in its annual periodicity, coriaceous texture and the associated white rot. Since it differs from all described species of *Lenzites,* it is proposed as a new species.

LOWEPORUS Wright

Mem. N.Y. Bot. Gard. 28 : 225, 1975.

Fruitbody resupinate to pileate, annual to perennial. Pileus surface when present grey, brown to fuscous, glabrous, mostly sulcate. Pore surface more or less of the same colour as pileus surface, tubes concolorous with pore surface. Context homogeneous and grey, brown to dark fuscous. Hyphal system di-to trimitic; generative hyphae with clamps; skeletal hyphae red, dextrinoid; binding hyphae present or absent. Cystidia none. Basidiospores hyaline to slightly tinted, ellipsoid to truncate, thin to thick-walled, variously dextrinoid.

Type species : *Polyporus lividus* Kalch. [=*Loweporus tephroporus* (Mont.) Ryv.]

Two species have been collected from Kerala.

KEY TO SPECIES

1. Fruitbody resupinate; pore surface ash grey to darker, pores 4-6 per mm; hyphal system trimitic.. *L. tephroporus*

1′. Fruitbody effused-reflexed to pileate; pore surface brown to violet brown, pores 6-8 per mm; hyphal system dimitic.. *L. fusco-purpureus*

LOWEPORUS FUSCO-PURPUREUS (Pers.) Ryv.

Prelim. Polyp. Fl. E. Africa p. 411, 1980. fig. 21. a-b; Pl. ., fig.

Polyporus fusco-purpureus Pers. In Gaudichand, Voy. Aut. Monde, Paris p.172, 1827.

P. calignosus Berk., J. Linn. Soc. 16 : 46, 1877.

Fomes calignosus (Berk.) Cke., Grevillea 14 : 20, 1885.

Polyporus calignosus Ces., Mycet. Borneo p. 5, 1889. Nomen illegit. Non Berk., 1877.

Pyropolyporus calignosus (Berk.) Imaz., Bull. Govt. For. Exp. Sta. Tokyo 57 : 110, 1952.

Phellinus calignosus (Berk.) G.H. Cunn., N.Z. Dep. Sci. Industr. Res. Bull 164 : 236, 1965.

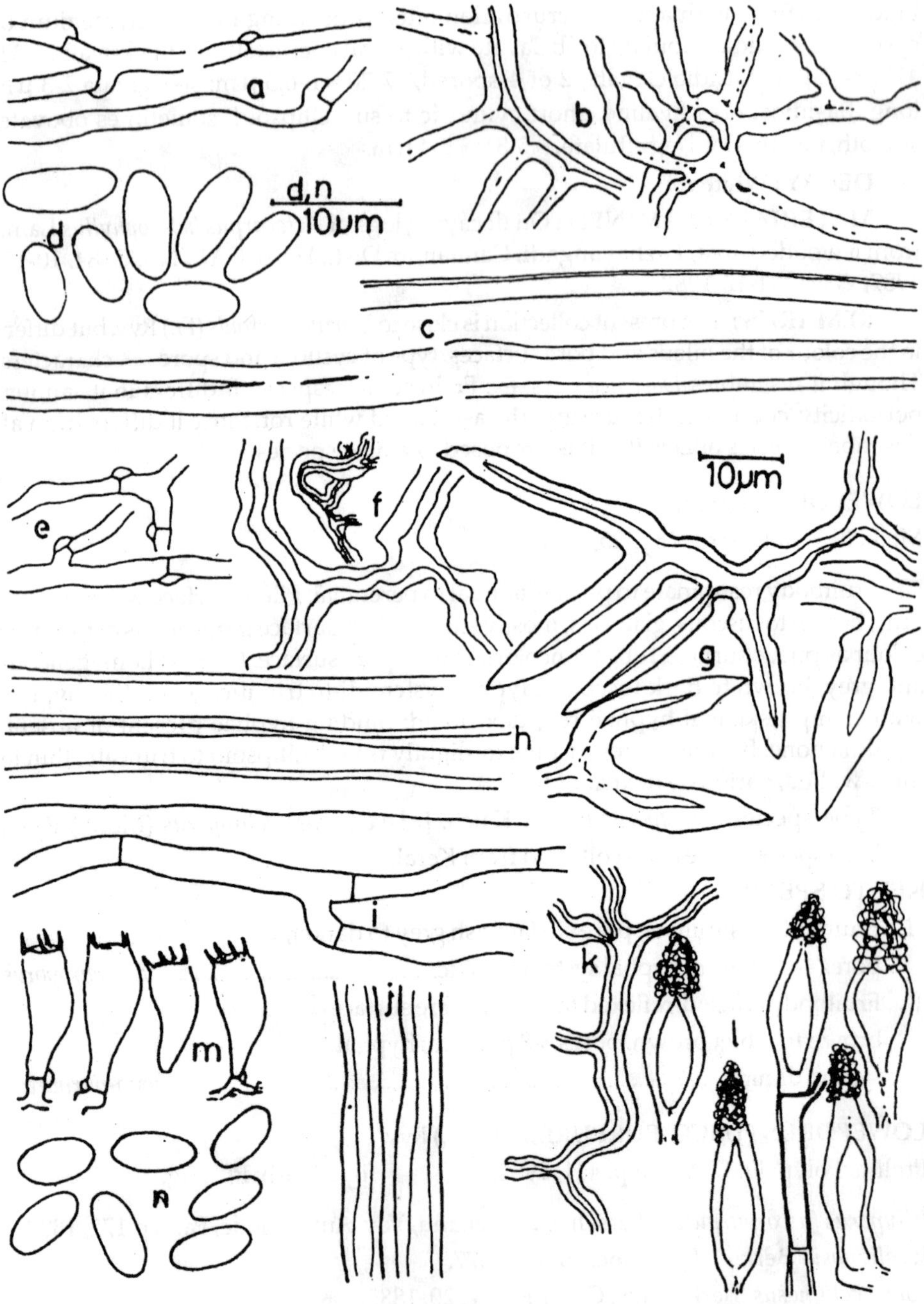

Fig. 20 : a-d *Lenzites acuta* : a-generative hyphae; b-binding hypahe; c-skeletal hyphae; d-basidiospores.

e-h *Lenzites betulina* : e-generative hyphae; f-binding hyphae; g-skeletal hyphae with sword-like side branchings; h-skeletal hyphae.

i-n *Lenzites* taxonomic species : i-generative hyphae ; j-skeletal hyphae; k-binding hyphae; l-cystidia; m-basidia; n-basidiospores.

Fruitbody annual to perennial, effused-reflexed to strictly pileate, sessile, attached with a broad to converging lateral base, solitary to imbricate, confluent; flabelliform to semi-circular to irregular, applanate to conchate, hard corky to somewhat woody; 5.5-13 × 3.5-9 × 1-2 cm. Pileus surface nutria (5F3) to almost black, often olivaceous brown (4E5) to greyish brown (6E3) or bronze (5E5), margin soot brown (5F5) to bronze (5E5); glabrous, uneven, concentrically striate and zonate, often sulcate, radially rugulose and faintly crusty; margin thick, rounded, faintly velutinate to glabrescent, incurved when dry. Pore surface snuff brown (5F6) to violet brown (11E4), greyish brown (11C2) to oak brown (5D6) towards margin, growth often not uniform; pores small and not visible to naked eye, absent near margin, 6-8 per mm; pore mouth 80-120μm broad; dissepiments 30-50 μm thick; pore tubes concolorous with the pore surface, stratified, 2-3 mm long in each stratum, brittle when dry. Context dark brown, reddish in KOH, 3-10 mm thick, absent between pore layers, homogeneous.

Hyphal system dimitic. Generative hyphae hyaline, thin-walled, branched, simple septate, infrequent, 1.75-2 μm in diam. Skeletal hyphae slightly pinkish or greyish pink, dark red in KOH, thick-walled, sparsely branched towards extremities, non-septate, usually cylindrical and straight, ends slowly tapering, 5-7. 5 μm in diam. Cystidia none. Basidia and basidiospores not observed.

DECAY : White fibrous rot.

SPECIMEN EXAMINED : On decaying forest logs, host not identified, Sholayar forest, Trichur Dist., Herb. CALI No. G96, 26-7-1984; Kadavakkunnu, Nadugani forests, Nilambur Forest Division, Nilambur, Malappuram Dist., Herb CALI No. G135, 13-11-1984.

GEOGRAPHICAL DISTRIBUTION : Philippines, Malaysia, Sri Lanka, India, New Guinea, Samoa, Australia, New Zealand, Africa.

REMARKS : In a study based on the herbarium specimens of persoon's collection, Ryvarden (1973) had first opined that *P. Fusco-purpureus* belongs to *Nigroporus* Murr. because of its dimitic hyphal system and clamped generative hyphae. In recent studies with African specimens (Ryvarden & Johansen, 1980) and Berkeley's collection (Ryvarden, 1984) he had included this species in *Loweporus.* Cunningham (1965) has included this species in *Phellinus.*

The first and only Indian record of this species as *Fomes calignosus* is from Andaman and Nicober Islands by Bakhi (1971) and it has smaller pores compared to those of the present collections. Description of this species form New Zealand (Cunningham, 1965) and East Africa (Ryvarden & Johansen, 1980) closely agree with the present collection except for the clamp connections in the generative hyphae reported by Ryvarden & Johnansen (1980).

LOWEPORUS TEPHROPORUS (Mont.) Ryv.

Prelim. Polyp. Fl. E. Africa p. 416, 1980.

fig. 21. c-g; Pl. fig.

Polyporus tephroporus Mont., Ann. Sci. Nat. 3,4 : 358, 1845.

P. surinamensis Mont., Ibid. 2, 20 : 363, 1842 nomen illegit. non Miq., 1839.

P. lividus Kalch. ex Cke., Grevillea 10 : 103, 1882. Non Cke., 1882.

P. luridus Kalch. ex Cke., Ibid. 12 : 21, 1883.

Fomes luridus (Kalch.) Sacc., Syll. Fung. 6 : 206, 1888.

Fomitoporia cubensis Murr., North Am. Fl.1 9 : 8, 1907.

Poria cinero-umbrina P. Henn., in Voeltzkow, Reise Ostafr. 3 : 19, 1908.

Fomes pallidus Petch, Ann. Roy. Bot. Gard. Paradeniya 6 : 135, 1916.

F. cinereus Torrend, Lloyd, Mycol. Letter 65 : 9, 1917.

F. olivaceus Lloyd, Ibid. 65 : 14, 1917.

F. longinquus Lloyd, Mycol. Writ. 7 : 1335, 1925.

Fuscoporia livida (Kalch.) G.H. Cunn., N.Z. Dep. Sci. Industr. Res. Bull. 164 : 216, 1965.

Fruitbody perennial, resupinate, adnexed to the host surface, spreading on the lower surface of branches or on crevices of vertically standing tree trunks, margin adpressed, hard woody in texture, 0.8-1.5 cm thick, spreading in sheets of about 10-30 × 10-30 cm or more. Pore surface ash grey (1B2) or lighter to greyish green (25B3), smoke grey (3C2) when dry, margin whitish, pale grey (IBI) when dry, uneven and rough; pores small and visible to naked eye, absent near, margin, round, almost regular, 4-6 per mm; pore mouth (110) 130-175 µm wide; dissepiments 5--100 µm thick; pore tubes stratified, fresh layer old silver (4E2), brownish grey (5D2) to concolorous with the pore surface when old, white mycelial strands irregularly distributed in between pore layers, 3-4 mm long in each stratum, tubes straight or oblique. Context old silver (4E2), about 1 mm thick in between pore layers. All tissues darkening in KOH.

Hyphal system trimitic. Generative hyphae hyaline, slightly thick-walled, branched, septate with clamps, cells shorter in dissepiments, frequent in context, 2-3. 5 µm in diam. Binding hyphae mostly restricted to the dissepiments but copious, hyaline, well branched, thick-walled with narrow lumen, non-septate, flexuous and twisted, 2-3 µm in diam. Skeletal hyphae olivaceous, thick-walled with broad and distinct lumen, usually unbranched or rarely branched near extremities, non-septate, cylindrical and straight, (2.5) 3. 5-5 in diam. Basidia sub-clavate to napiform, easily collapsing, 4-spored, 8-10 × 3.75-5 µm; sterigmata slender and long, up to 3 µm long. Basidiospores hyaline, slightly thick-walled, oblong to sub-ovate, truncate in some; often with 1-2 oil globules, slightly dextrinoid, 5-6 × 3.75-4 µm.

DECAY : White spongy rot of heartwood.

SPECIMEN EXAMINED : On decaying trunks of *Swietenia macrophylla* King (MELIACEAE), Kuppadi, Sultan Battery, Wayanaad Dist., Herb. CALI No. G191,

19-6-1985; on living trunks of *Melia azadirachta.* Linn. (MELIACEAE), Poothole, Trichur Dist., Herb. CALI No. G328, 17-8-21986; on unidentified decaying strumps, Noolpuzha, Muthanga Reserve Forest, Wyanaad Dist., Herb. CALI No. G316, 22-6-1986.

GEOGRAPHICAL DISTRIBUTION : East Africa, French Antilles, French Guiana, Argentina, North America, Australia, India.

REMARKS : This polypore has earlier been reported form Calcutta by Bakshi (1971) as *Fomes lividus* (Kalch.) Sacc.

MICROPORELLUS Murr.

Bull. Torrey Bot. Club. 32 : 483, 1905.

Fruibody annual, stipitate, flabelliform to spathulate, rather thin, coriaceous when fresh, rather hard when dry, stipe central or lateral, often somewhat reduced to a long tapering base. Pileus surface tomentose to smooth, usually distinctly zonate, ochraceous to bay; pores small. Context white to ochraceous. Hyphal system dimitic; generative hyphae with or without clamps; skeletal hyphae present in trama, absent or present in the context. Cystidia none. Basidiospores globose to sub-globose, hyaline, smooth, thin-walled and nonamyloid. On deciduous wood. Pantropical genus.

Type species : *Polyporus dealbatus* Berk. & Curt. [=*Microporellus dealbatus* (Berk. & Curt.) Murr.]

Only one species has been recorded from Kerala.

MICROPORELLUS OBOVATUS : (Jungh.) Ryv.

Norw. J. Bot. 19 : 232, 1972. fig. 21-h-k; Pl., Fig.

Polyporus obovatus Jungh., Verh. Batav. Gonootsch.17 : 65, 1838.

P. lacerus Jungh., Verh. Batav. Genootsch. 17: 65, 1838.

P. delitatus Berk., Lond. J. Bot. 6 : 499, 1847, nomen illegit. Non Lev., 1844.

P. umbilicatus Berk., Hook. J. Bot. 3 : 79, 1851 nomen illegit. non Jungh. 1838.

P. mutabilis Berk. & Curt., Ann. Mag. Nat. Hist. 2, 12 : 433, 1853.

P. polygrammus Berk. & Curt., J. Linn. Soc. Bot. 10 : 307, 1869.

P. ravenalii Berk. & Curt., Grevillea 1 : 38, 1872.

P. gallinaceus Berk., &. Cke., J. Linn. Soc. Bot. 15 : 379, 1876.

P. rasipes Berk., Ibid. 16 : 49, 1878.

P. caryophllaceus Berk. & Curt., Gervillea 15 : 22, 1887.

P. tigrinus Rostr., Bot. Tidsskr. 24 : 359, 1902.

Fruitbody annual, solitary or in small gregarious group, laterally stipitate, rarely with a reduced stipe, confluent, flabelliform, reniform or spathulate, coriaceous while fresh, tough and reflexed when dry; 2-5.5 × 2-5 × 0.2-0.4 cm. Pileus surface butter yellow (4A5) with concentric striations of mustard brown (5E6), sometimes striations of greyish brown (10F3), liver brown (8F6) and butter yellow (4A5) in alternate bands, pattern of striation varying from fruitbody to fruitbody; even and smooth, faintly vellutinate while young, often glabrous when mature; margin thin, entire, incurved when dry. Stipe light brown (5D8) or yellowish brown (5E8), finely velutinate while young, becoming glabrous when old, -2.5 cm long, 0.3-0.6 cm thick, cylindrical, solid, stipe-tissue cream (4A3). Pore surface white to light yellow (4A4) to greyish white (1B2), brownish orange (5B3) when dry, smooth and even; pores never decurrent, minute and not visible to unaided eye, absent near margin, round, regular, 7-8(9) per mm; pore mouth 50-65 μm wide; dissepiments 25-40(50) μm thick; pore tubes white with a yellowish tinge or yellowish white (3A2), up to 1.5 mm long, straight, sunk into equal depths. Context white to cream (4A3), up to 2 mm thick.

Hyphal system dimitic. Generative hyphae hyaline, thin-walled to slightly thick-walled, either simple-septate or with clamps, sparsely branched, common on context, 2.5-5 μm in diam. Skeletal hyphae hyaline, thick-walled with narrow but distinct lumen, unbranched, non-septate, cylindrical, and equal in width, 4-6 μm in diam. Basidia clavate to somewhat utriform, 2 or 4-spored, 9.75-12 × 2.25-6 μm; sterigmata up to 1.5 μm long. Cystidia absent. Basidiospores hyaline, ovoid to sub-globose, thin-walled, smooth, 0 to 1-guttulate, nonamyloid, 4.5-5.5 × 3-4 μm.

DECAY : White fibrous rot.

SPECIMEN EXAMINED : On unidentified forest logs, Kanjirakadavu, Nilambur Forest Division, Malappuram Dist., Herb. CALI No. G92, 16-9-1983; Malakkapra, Sholayar forests, Trichur Dist., Herb. CALI No. G180, G284, 10-6-1985; Vellarimala, Chembra Peak, Meppadi, Wyanaad Dist., Herb. CALI No. G240, 11-9-1985.

GEOGRAPHICAL DISTRIBUTION : Pantropical species; tropical Africa, French Antilles and Guiana, India.

REMARKS : This is the first record on the occurrence of *M. Obovatus* in India. The predominently spathulate, laterally stipitate, thin fruitbodies, minute pores, mostly subglobose basidiospores and dimitic hyphal system are characteristic of this species. According to David & Rajchenberg (1985) the macro-and micromorphological features of this species do not match with the definitions of the genus *Microporellus* Murr., and the species should be excluded from the genus although provisionally they have employed the binomial *M. Obovatus*. Genus *Microporellus* resembles *Favolus* in general form, but the latter has much larger pores and spores and simple-septate generative hyphae.

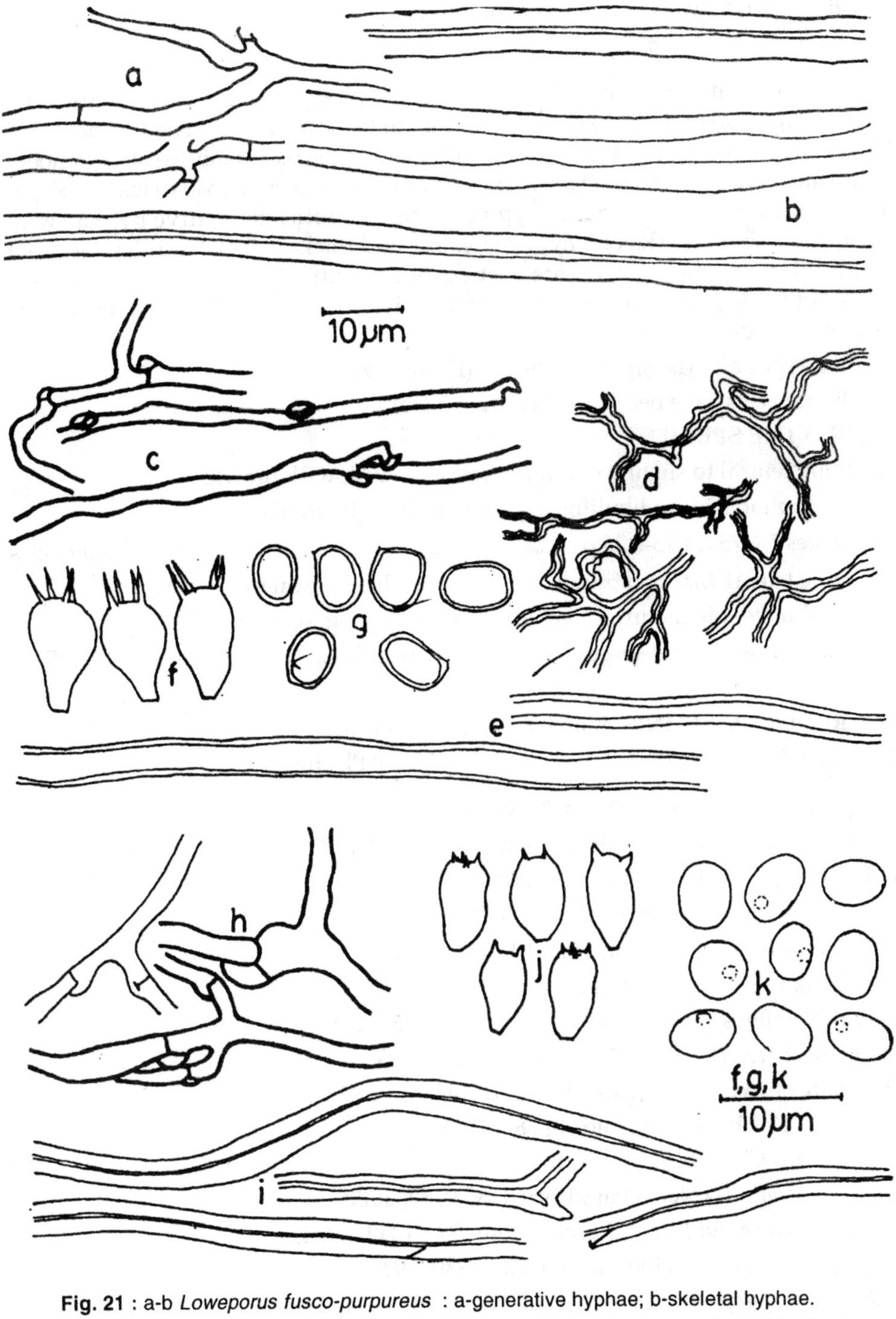

Fig. 21 : a-b *Loweporus fusco-purpureus* : a-generative hyphae; b-skeletal hyphae.
c-g *Loweporus tephroporus* : c-generative hyphae; d-binding hyphae; e-skeletal hyphae; f-basidia; g-basidiospores.
h-k *Microporellus obovatus* : h-generative hyphae; i-skeletal hyphae; j-basidia; k-basidiospores.

MICROPORUS Beauv. ex Kuntze emend Pat.
Rev. Gen. Pl. 3 : 494, 1898.

Fruitbody annual, centrally or laterally stripitate, circular, infundibuliform, flabelliform or spathulate. Pileus surface smooth to hirsute, often zoned. Stipe lateral or central, round and usually with expanded foot at the base, white to black, smooth or hirsute. Pore surface white to cream, pores round and entire, very small, 5-10 per mm; context white and tough. Hyphal system trimitic; generative hyphae with clamps; binding and skeletal hyphae hyaline and thick-walled. Cystidia none, but coralloid dichophytic elements occur along the dissepiments. Basidiospores allantoid to elliptical, smooth, thin-walled, hyaline, nonamyloid. On deciduous woods. Tropical genus.

Type species : *Microporus xanthopus* (Fr.) Kuntze

Two species have been recorded from Kerala.

KEY TO THE SPECIES

1. Stipe central to slightly excentric, yellow to yellowish brown; sporophores infundibuliform; pileus surface glabrous; spores 5-6.75 × 1.75-2.25 μm.. *M. xanthopus*

1′. Stipe lateral, blackish brown; sporophores flabelliform to spathulate; pileus surface velutinate while young, seldom glabrous when old; spores 3.5-4 × 1.5-2 μm.. *M. affinis.*

MICROPORUS AFFINIS (Blume & Nees ex Fr.) Kuntze.
Rev. Gen. Pl. 3 : 494, 1898. Fig. 22. f-j; Pl. , fig.

Polyporus affinis Blume & Nees, Nova Acta 13 : 18, 1826.
P. affinis Blume & Nees ex Fr., Elench. Fung. P. 126, 1826.
P. luteus Nees, Nova Acta 13 : 16, 1826.
P. flabelliformis Kl., Linnaea 8 : 483, 1833.
P. crenatus Berk., Ann. Nat. Hist. 10 : 372, 1843.
P. microloma Lev., Ann. Sci. Nat. 3, 2 : 183, 1844.
Polystictus affinis (Nees) Fr., Nova Acta 3, 1 : 75, 1851.
Polyporus trizonatus Cke., Grevillea 12 : 17, 1883.
Polystictus flabelliformis (Kl.) Cke., Ibid. 14 : 78, 1886.
P. glirinus Kalch. ex Cke., Ibid. 14 : 83, 1886.
P. trizonatus Cke., Ibid. 14 : 84, 1886.
Microporus affinis (Nees) Pat., Essai Tax. P. 83, 1900.
Polyporus krakataui Lloyd, Mycol. Notes 71 : 1242, 1924.
Polystictus ruforigidus Lloyd, Ibid. 73 : 1330. 1924.
Polystictus roseoporus Lloyd, Mycol. Notes 73 : 1331, 1924.
Coriolus flabelliformis (Kl.) G.H. Cunn., Proc. Linn. Soc. N.S.W. 75 : 226, 1950.

Fruitbody annual, pileate, sub-stipitate to stipitate, solitary or in gregarious groups, imbricate and connate; flabelliform to spathulate or semi-circular, sometimes reniform or dimidiate, applanate, coriaceous while fresh, rigid when dry; 2.5-9.5 × 2-9 × 0.2-0.4 cm. Pileus surface brownish yellow (5C8) to reddish golden (6C7) or brownish orange (6C8), often dark brown (9F6), reddish brown (9E6), rose wood (9D5), terra cota (7D7), caramel (6C6), Pompeian yellow (5C6) or apricot (5B6) while fresh, colour varying from fruitbody to fruitbody, concentrically zonate, smooth, depressed (concave) and finely velutinate near stipe, elsewhere glabrous, shiny; margin entire, smooth, lateral, rudimentary or well developed. Stipe Pompeian yellow (5C6) to agate (7E8) often black; finely velutinate, hairs more concentrated towards the expanding base, 0.5-2.5 cn long and 3-5 mm thick; stipe-tissue white to cream-coloured, solid. Pore surface white to cream (4A3) while fresh, often reddish brown (8E8) near margin, cream (4A3) or with a brownish tinge when old and dry, even and smooth; pores never decurrent, absent near margin, minute and not visible to naked eye, round to angular, (7) 8-10 per mm; pore mouth 55-70(90) μm wide; dissepiments 35-75(95) μm thick; pore tubes concolorous with the pore surface, straight, sunk into equal depths, 1-1.5 mm long. Context white while fresh, yellowish white (4A2) when dry; homogeneous, 1-2.5 mm thick.

Hyphal system trimitic. Generative hyphae hyaline, slightly thick-walled, branched, nodose-septate, sparse in context, 2.25-3 μm in diam. Binding hyphae hyaline to pale-coloured, thick-walled, well branched, nonseptate, 2-3(4.5) μm in diam. Skeletal hyphae hyaline to pale-coloured, thick-walled with obliterate lumen or solid, unbranched, straight and slightly flexuous, common in context, 2-25-3.75 (7.5) μm in diam. Basidia sub-clavate, 2 or 4-spored, 9-12 × 3.75-5.25 μm; sterigmata straight, 2.25-3 μm long. Cystidia none. Basidioles numerous, sub-clavate, 7-9 × 3. 5-5 μm. Basidiospores hyaline, broadly ellipsoid, thin-walled, smooth, 1 to 2-guttulate, nonamyloid, 3.5-4 × 1.5-2 μm; spore print white.

DECAY : White rot.

SPECIMEN EXAMINED : On decaying branches of *Terminalia* sp. (COMBRETACEAE), Chooralmala, Meppadi, Wyanaad Dist., Herb. CALI No. G113, 21-9-1984; on decaying longs of *Elaeocarpus tuberculatus* Roxb. (ELAEOCARPACEAE), Kanoth Range, Wyanaad Dist., Herb. FRI No. 4340, K. Bagchee, 25-1-1946; on unidentified forest logs, Panthenthode, Silent Valley National Park, Palghat Dist., Herb. CALI No. G33, 7-9-1983; Kerala-Mysore State boundary, Kasaragod Dist., Herb. CALI No. G41, 8-12-1983; Kaduvakkunnu, Nadugani forests, Nilambur, Malappuram Dist., Herb. CALI No. G46, G47, G51, 28-1-1984; G128, G134, 13-11-1984; Vettilapara, Sholayar forests, Trichur Dist., Herb. CALI No. G77, 26-7-1984; Vengoli Peak, teak Plantation Division, Parambiculam, Palghat Dist., Herb. CALI No. G212, 11-7-1985; Karadimala Brahmagiri Ranges, Tirunelli, Wyanaad Dist., Herb CALI No. G262, G263, 2-10-1985; Noolpuzha, Muthanga Reserve Forests, Wyanaad Dist., Herb. CALI No. G138, 22-6-1986; Peria Reserve Forests, Wyanaad Dist., Herb. FRI No. 4338, K. Bagchee, 24-1-1946; Kanijirakadavu block, Amarambalam range, Nilambur, Malappuram Dist., Herb. FRI No. 4238, K. Bagchee, 26-1-1946; Alleppey Dist., Herb. FRI No. G503, IFS students, 30-4-1965.

GEOGRAPHICAL DISTRIBUTION : Malaysia, Java, India, Sri Lanka, Australia, New Zealand, Fiji, Mauritius, East and West Africa, Central America, New Caledonia, New Guinea, West Indies.

REMARKS : This is a widely distributed species in the forests of Kerala showing variations in colour and shape of the sporophores. Ths variability of *M. affinis* has already been mentioned by Ryvarden (1976b) and Ryvarden & Johansen (1980). Because of this, it has been described under different names and there is still disagreement among workers (Cunningham, 1965) about the specimens included under *M. affinis.* Bakshi (1971) has reported this species as *Polyporus flabelliformis* from Assam, West Bengal, South India and South Andamans.

MICROPORUS XANTHOPUS (Fr.) Kuntze

Rev. Gen. Pl. 3: 494, 1898. fig. 22. f-j; Pl. fig.

Polyporus xanthopus Fr., Syst. Mycol. 1: 350, 1821.

P. saccatus Pers. in Gaudichaud, C. Voy. Aut. Monde. Paris p. 169, 1826.

P. perula (P. Beauv.) Fr., Epicrisis p. 437, 1838.

Polystictus xanthopus Fr. Nova Acta 3, 1: 74, 1851.

Polyporus cupreo-nitans Kalch., in thuem., Mycol Univ. exsicc. p. 1702, 1881.

Coriolus xanthopus (Fr.) G. H. Cunn., Proc. Linn. Soc. N.S.W. 75: 247, 1950.

Fruitbody annual, stipitate, solitary or in groups, confluent, infundibuliform, coriaceous, petal-like, thin and flexible; 3.5-9 cm. in diam., up to 1.0 mm thick. Pileus surface violet brown (10E5) to pastel red (10E4) and its allied shades in concentric bands, often violet brown (10F3) at the centre to black towards the margin; radially wrinkled, occasionally splitting, glabrous, shiny; margin pale yellow (3A3), thin and entire. Stipe central or slightly excentric light orange (5A4) to light yellow (3A5), glabrous, base expanding, 0.5-2.5 cm long, 2-4 mm thick, solid, stipe-tissue cream (4A2) to alabaster ((5B2), smooth, even; pores never decurrent, extending almost to the margin, minute and not visible to the naked eye, round, regular, 8-10 per mm; pore mouth 50-65 μm wide; dissepiments 50-75(100) μm thick; pore tubes concolorous with the pore surface, shallow, up to 0.7 mm long, sunk into equal depths. Context white, up to 1.0 mm thick, homogeneous.

Hyphal system trimitic. Generative hyphae hyaline, thin-walled, branched, septate with clamps, 1. 5-2. 5 μm in diam. common in dissepiments and growing areas. Binding hyphae pale-coloured, thick-walled with narrow or obliterating lumen, richly branched, interwoven, non-septate, 2.5-3.75 μm in diam. Skeletal hyphae pale-coloured, thick-walled with narrow or obliterating lumen, unbranched, 3.5-6.25 μm in diam., dominating. Basidia clavate, 4-spored. 10-12 × 3.75-4.25 μm; sterigmata straight up to 2.5 μm long. Cystidia none. Basidiospores cylindrical, hyaline, smooth, thin-walled, eguttulate, nonamyloid, 5-6.75 × 1.75-2.25 μm. Hymenial elements collapsing into honey comb-line structure.

DECAY : White rot.

SPECIMEN EXAMINED : On decaying branchlets of *Careya arborea* Roxb. (LECYTHIDACEAE), Muthanga Reserve Forests, Noolpuzha, Wyanaad Dist., Herb. CALI No. G312, 22-6-1986; on dead branches of *Toona ciliata* Roemr (MELIACEAE), Kanoth range, Wyanaad Dist., Herb. Fri No. 4313, K. Bagchee, 24-1-1946; on identified forest tree branchlets, Paringalkuttu, Sholayar forests, Trichur Dist., Herb. CALI No. G10, 28-6-1983; Panthenthode, Silent Valley National Park, Palghat Dist., Herb. CALI No. G28, 30-5-1983; G58, 28-2-1984; Vazhachal, Sholayar forest, Trichur Dist., Herb. CALI No. G85, 26-7-1984; Kanjirakadavu, Amarambalam Range, Nilambur, Malappuram Dist., Herb. CALI No. G93, 19-9-1983; Mukkali, Silent Valley National Park, Palghat Dist., Herb. CALI No. G107, 7-9-1983; Malakkapara, Sholayar Forests, Trichur Dist., Herb. CALI No. G1178, 10-6-1985; Vellarimala, Chembra Peak, Meppadi, Wyanaad Dist., Herb. CALI No. G239, 11-9-1985; Parambiculum. Teak Plantation Division, Palghat Dist., Herb. CALI No. G276, 16-1-1986.

GEOGRAPHICAL DISTRIBUTION : Africa, South America, New Guinea, Pacific Islands, Australia, New Zealand, Philippines, China, Singapore, India.

REMARKS : *M. xanthopus* is common in forests throughout Kerala. The shape and colour of sporophores and much bigger spores distinguish it from *M. affinis* which is also quite widely distributed in the forests of Kerala. This species has been reported earlier from India (Bose, 1946; Bakshi, 1971) as well as from Kerala (Sharma et al., 1985).

NIGROPORUS Murr.

Bull. Torrey Bot. Club. 32 : 361, 1905.

Fruitbody annual to perennial, pileate to resupinate. Pileus surface greyish blue, vinaceous brown to pink or violet, scrupose to glabrous, azonate to concentrically zonate. Pore surface concolorous with the pileus surface, pores usually small, entire, round to angular. Context vinaceous brown to pink or purplish. Hyphal system dimitic; generative hyphae with clamps, skeletal hyphae fuliginous brown, thick-walled to solid. Cystidia none. Basidiospores mostly small, to 5μm, hyaline, smooth, thin-walled, allantoid to broadly ellipsoid, nonamyloid. On deciduous wood. Pantropical genus.

Type species : *Nigroporus vinosus* (Berk.) Murr.

Two species have been collected from Kerala.

KEY TO SPECIES

1. Pileate species; pores 8-10 per mm; causing white rot.............................. *L. Vinosus*

1′. Fruitbody resupinate; pores 4-6(7) per mm; causing brown rot................ *L. Niger*

NIGROPORUS NIGER (Berk.) Ryv.

Norw. J. Bot. 24 : 223, 1977.

fig. 22. o-q; Pl. 29, fig.

Polyporus niger Berk., Lond. J. Bot 4 : 304, 1845.

Melanoporia niger (Berk.) Murr., North Am. Fl. 9 : 14, 1907.

Fruitbody perennial, broadly effused, adnate, inseparable from host surface, woody hard while fresh and dry; 2-10 X 1.5-10 cm or more in size, 0.5-0.8 cm thick. Pore surface pale grey (IBI) to black towards older region; pores small and not visible to naked eye, 4-6(7) per mm; pore mouth 100-180 μm wide; dissepiments 40-80 μm thick; pore tubes concolorous with the pore surface, inside walls white, stratified, 2-3 mm long, usually straight, sunk into equal depths. Context dark grey, a very thin layer in between pore strata.

Hyphal system dimitic. Generative hyphae hyaline, thin-walled, branched, septate with clamps, not frequent, 2.5-3.25 μm in diam. Skeletal hyphae slightly blackish olive, thick-walled to solid, sparsely branched towards extremities, cylindrical and straight, 5-7.5 μm in diam. Cystidia and setae absent. Basidia clavate, usually 4-spored, 12-13.75 × 3-3.75 μm; sterigmata up to 1.5 μm long. Basidiospores not observed.

DECAY : Brown cuboidal rot.

SPECIMEN EXAMINED : On decaying logs of *Tectona grandis* Linn. f. (VERBENACEAE), Kanjirakadavu, Amarambalam Range, Nilambur, Malappuram Dist., Herb. CALI No. G142, 15-2-1985; on unidentified forest logs, Kanjirakadavu Nilambur, Malappuram Dist., Herb. CALI No. G143, 15-2-1985.

GEOGRAPHICAL DISTRIBUTION : North America, India.

REMARKS : It was Ryvarden (1977) who made the new combination *Nigroporus niger* during the type studies of species described by M.J. Berkeley. He described a black resupinate fruitbody with hyaline and oval spores.

This is the first report on the occurrence of *N. niger* in India. This polypore is characterised by its greyish black pileate to resupinate fruitbody, regular pores, dimitic hyphal system with coloured skeletal hyphae, clamped generative hyphae and associated brown rot.

NIGROPORUS VINOSUS (Berk.) Murr.

Bull. Torrey Bot. Club 32 : 361, 1905. fig. 22. k-n; Pl., fig.

Polyporus vinosus Berk., Ann. Mag. Nat. Hist. 2, 9 : 195, 1852

P. tristis Lev., Ibid. 3, 5 : 126, 1846, non Pers, 1825.

P. badius Jungh. In Moritzi Syst. Verz. Zollinger p. 120, 1846, nomen illegit, non Berk., 1841; non Pers. ex Schw., 1832.

P. xerophyllaceus Berk., Hook. J. Bot. 8 : 200, 1856.

Polystictus xerophyllaceus (Berk) Cke., Grevillea 14 : 86, 1886.

P. Vinosus (Berk.) Cke., Ibid. 15 : 51, 1886.

Trametes tristis (Lev.) Cke., Ibid. 15 : 60, 1886.

Polystictus mollerianus Sacc., Berl. & Roum., Rev. Mycol. 11, 14 : 202, 1889.

Microporus mollerianus (Sacc., Berl. & Roum.) Kuntze, rev. Gen. Pl. 3, 3 : 496, 1898.

Microporus vinosus (Berk.) Kuntze, Rev. Gen. Pl. 3, 3 : 497, 1898.

M. xerophyllaceus (Berk.) Kuntze, Ibid. 3, 3 : 497, 1898.

Coriolus vinosus (Berk.) Pat., Essai Tax. P. 94. 1900.

C. tristis (Lev.) Pat., Ibid. P. 94, 1900.

Polystictus badius (Jungh. In Moritzi) Bres., Ann. Mycol. 10 : 504, 1912.

Fomitopsis vinosa (Berk.) Imaz., Bull. Govt. For. Exp. Sta. Tokyo. 57 : 111, 1952.

Fruitbody annual or biennial, pileate, sessile, attached with a narrow lateral base, imbricate, confluent, flabelliform to spathulate, sometimes semi-circular, thin, coriaceous while fresh, remaining flexible when dry; 2-6.5 × 2-4 × 0.1-0.2 cm. Pileus surface brownish orange (7C4), dark brown (7F4), Madeira (8E5), or greyish brown (8F3), black towards the base, white with a pinkish tinge along the margin while fresh soon fading to greyish brown (8F3), glabrous, concentrically zonate, zonations closer towards margin; margin thin, smooth, entire. Pore surface pinkish white (10A2) when collected soon fading to brownish grey (7C2) or red-haired (6C4), pores minute and not visible to naked eye, round, 8-10(11) per mm; pore mouth 60-80 μm wide; dissepiments 35-65 μm thick; pore tubes concolorous with the pore surface, rarely stratified, up to 0.5 mm long, straight, sunk into equal depths. Context reddish brown (8D4) to Somalis (7E5); homogeneous, up to 1.5 mm thick.

Hyphal system dimitic. Generative hyphae hyaline to light grey (or brown), slightly thick-walled, branched, usually nodose-septate but sometimes simple-septate, 3-3.75 μm in diam., very common in context and dissepiments. Skeletal hyphae brown, thick-walled with narrow but distinct lumen, unbranched, non-septate, cylindrical and straight, 3.25-4.5 μm in diam. Basidia short, clavate, usually 4-spored, 7.5-12 × 3.5 μm; sterigmata up to 1.5 μm long, straight. Cystidia absent. Hyphal pegs present. Basidiospores hyaline almost cylindrical to slightly allantoid, thin-walled, smooth, 0 to 2 guttulate, nonamyloid, 3.5-4.5 × 1.5-2 μm.

DECAY : White rot.

SPECIMEN EXAMINED : On decaying forest logs of *Palaquium ellipticum* (Dalz.) Bail. (SAPOTACEAE), Perumpara (W), Malakkapara, Sholar forests, Trichur Dist., Herb. CALI No. G354, 30-8-1986; on unidentified forest logs, Kaduvakkunnu, Nadugani forests, Nilambur, Malappuram Dist., Herb. CALI No. G48, 28-1-1984; near abandoned Dam site, Silent Valley National Park, Palghat Dist., Herb. CALI No. G223, 15-9-1985; Mundakai forests, Meppadi, Wyanaad Dist., Herb. CALI No. G244, 12-9-1985.

GEOGRAPHICAL DISTRIBUTION : Throughout the tropical world.

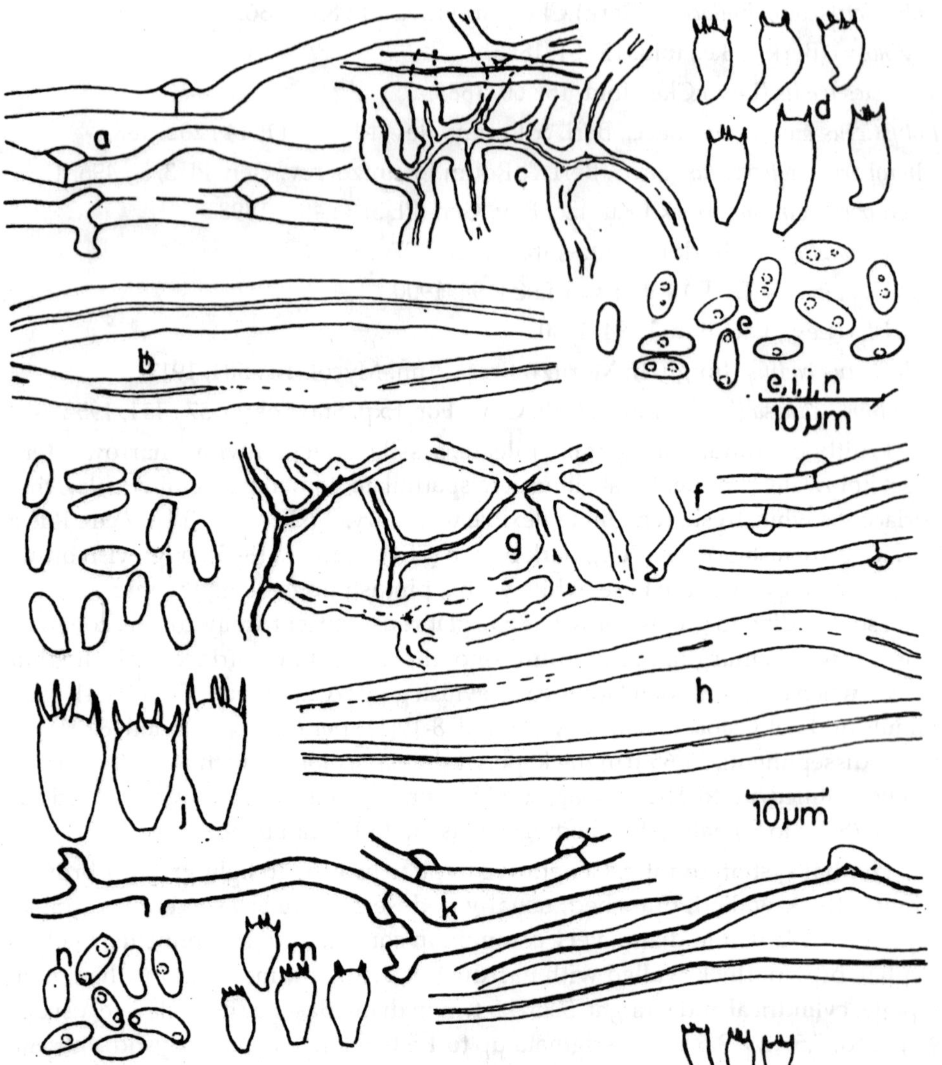

Fig. 22 : a-e *Microporus affinis* : a- generative hyphae; b- skeletal hyphae; c- binding hyphae; d- basidia; e- basidiospores.

f-j *Microporus xanthopus* : f- generative hyphae; g- binding hyphae; h—skeletal hyphae ; i - basidiospores; j - basidia.

k - n *Nigrocorus vinosus* : k- generative hyphae; l- skeletal hyphae; m - basidia; n- basidiospores.

o - q *Nigroporus niger* : o - generative hyphae; p - skeletal hyphae; q - basidia.

REMARKS : Dark vinaceous pileus surface, minute pores and associated white rot are most characteristics of *N. vinosus*. Bose (1934, 1937a) and Bakshi (1971) described this species from India, the latter describing it as having a monomitic hyphal system. The present collection agrees closely with the description of *N. vinous* from tropical America (Murrill, 1915) and East Africa (Ryvarden & Johansen, 1980) But collection from Trinidad and Tobago by Fidalgo & Fidalgo (1967) have wider pores and thicker dissepiments.

OXYPORUS Donk :
Meddel. Bot. Mus. Univ. Utrecht 9 : 212, 1933.

Fruitbody annual to perennial, pileate to resupinate, when pileate mostly broadly attached. Pore surface white, cream to straw-coloured; pores entire, medium to small; tubes non-stratified or with distinct zones. Hyphal system monomitic, generative hyphae without clamps, thin-to thick-walled. Cystidia present, ventricose to tubular and smooth or apically encrusted or more oblong and encrusted in the upper part. Basidiospores globose to broadly ellipsoid. On deciduous and coniferous wood. Cosmopolitan genus.

Type species : *Oxyporous populinus* (Fr.) Donk.

Two species have been recorded from Kerala.

KEY TO SPECIES

1. Fruitbody annual; cystidia cylindro-clavate, up to 37.5 X 10 μm. Spores hyaline, cylindrical to oblong-ellipsoid. 6-8(11) × 3-4(5) μm.......... *O. cervino-gilvus*

1′. Fruitbody bienial to perennial; cystidia long, cylindrical. often septate, up to 75 × 12.5 μm; spores hyaline, ovoid to broadly ellipsoid, (9) 11-13 X 7-8 μm.......... *O. mollissimus*

OXYPORUS CERVINO-GILVUS (Jungh.) Ryv.
Norw. J. Bot. 20 : 3, 1973. fig. 23; a-d. Pl., fig.

Polyporus cervino-gilvus Jungh., Verh. Batav. Genootsch. 17 : 45, 1838.
Trametes dermatoides Lev., Ann. Sci. Nat. 3, 2 : 196, 1844.
Polyporus dermatoides (Lev.) Lev., Voy. La. Bonike l : 180, 1846.
Polystictus cervino-gilvus (Jungh.) Fr., Nova Acta 3, 1 : 94, 1851.
Polyporus personatus Berk. & Br., J. Linn. Soc. 14 : 51, 1873.
P. peradeniae Berk. & Br., Ibid. 14 : 51, 1873
P. diversiporus Berk. & Br., Ibid. 14 : 53, 1873.
P. chrysoleucus Kalch. ex Thuem., Grevillea 4 : 72, 1875.
Polystictus chrysoleucus (Kalch.) Cke., Ibid. 14 : 82, 1886.
P. personatus (Berk. & Br.) Cke., Ibid. 14 : 84, 1886.

P. peradeniae (Berk. & Br.) Cke. Ibid. 14 : 84, 1886

Poria diversipora (Berk. & Br.) Cke., ibid. 14 : 113, 1886.

Trametes peradeniae (Berk. & Br.) Pat., Essai Tax p. 93, 1900.

Coriolus peradeniae (Berk. & Br.) G.H. Cunn., Proc. Linn. Soc. N.S.W. 75 : 236, 1950.

Trichaptum cervino-gilvus (Jungh.) G.H. Cunn., N.Z. Dep. Sci. Industr. Res. Bull. 165 : 96, 1965.

Fruitbody epixylous, annual, resupinate to effused-reflexed, inseparable from the wood, up to 2mm thick, coriaceous and flexible while fresh, slightly tough or rarely brittle when dry. Pileus surface snuff brown (5F6) to mustard brown (5E6), pale yellow (1A3) towards periphery; matted adpressed-tomentose, uneven; margin thin, irregular and wavy. Pore surface mustard yellow (3B6) to wheat (4B5) or pale yellow (1A3) with certain patches of light orange (5A4), margin corn (4B5) or golden wheat (5B5) coloured; pores large and visible to unaided eye, not extending to the margin, angular to round or ovate, regular, 1-2 per mm; dissepiments 150-210 µm thick; pore tubes concolorous with the pore surface, up to 1 mm long. Context linoleum brown (5E7) or darker, less than 1 mm thick.

Hyphal system monomtic. Generative hypahe hyaline to cream, thin to moderately thick walled with distinct lumen, branched and simple-septate, cylindrical and almost equal in width, 3-5 µm in diam. Cystidia hyaline, thick-walled, ventricose to clavate, encrustation usually apical, 19.5-28.5(37.5) × 6-7.5(10) µm. Basidia clavate, 4-spored, 15-18 × 4-6.5 µm; sterigmata 3-4 µm long. Basidiospores hyaline, cylindrical to oblong-ellipsoid, thin-walled, smooth, eguttulate, nonamyloid, 6-8(11) × 3-4(5) µm.

DECAY : White rot.

SPECIMEN EXAMINED : On decaying logs of *Tectona gradis* Linn. (VERBENACEAE), Divisional Forest Office premises, Nilambur, Malappuram Dist., Herb. CALI No. G145, G149, 7-3-1985; on decaying logs of *Artocarpus heterophyllus* Lam. (MORACEAE), Pazhayangadi, Cannanore Dist., Herb. CALI G-386, 31-12-1987; on unidentified forest logs, Vellarimala, Chembra Peak, Meppadi, Wyanaad Dist., Herb. CALI No. G236, 11-9-1985; on forest logs, Herb. FRI No. 4250, K. Bagchee, 28-1-1946.

GEOGRAPHICAL DISTRIBUTION : India, Sri Lanka, Java, Philippines, China, Fiji, Australia, East Africa.

REMARKS : The present collection agrees with the description of this fungus by earlier Indian workers (Bose, 1937a; Bakshi, 1971). *O. Cervino-gilvus* is characterised by brownish, tomentose pileus surface, large pores, monomitic hyphal system and oblong-ellipsoidal spores. It is different from other species of *Oxyporus* in the colour of the pileus, which is cinnamon to mustard brown while it is straw-coloured in rest of the species.

OXYPORUS MOLLISSIMUS (Pat.) Reid.
Microscopy 32 : 456, 1975. fig. 23. e-h.; Pl., fig.

Polyporus mollissimus Pat., J. Bot. Paris 1 : 340, 1897.
Xanthochrous berneiri Har. & Pat., Ibid. 17 : 9, 1903.
Spongipellis stramineus Pat., Bull. Soc. Mycol. Fr. 24 : 166, 1908.
Polyporous stramineus (Pat.) Sacc. & Trott. Syll. Fung. 21 : 166, 1908.
Poria xylina Lloyd, Mycol. Writ. 5 : 630, 1917.
Trametes straminea (Pat.) Lloyd, Mycol. Letter 69 : 13, 1919.
T. irpicoides Pilat, Bull. Soc. Mycol. Fr. 52 : 311, 1937.
Polyporus irpicoides Bond. In Pilat, Atl. Polyp. P. 305, 1939.
Trametes heteromorpha (Fr.) Bres. Var. *Irpicoides* (Pilat) Pilat, Atl. Plyp. P. 305, 1939.
Leucophellinus irpicoides (Pilat) Bond. & Sing., Mycologia 36 : 68, 1944.
Oxyflavus mollissimus (Pat.) Ryv., Norw. J. Bot. 20 : 4, 1973.
Leucophellinus mollissimus (Pat.) Parm., Eesti Nsv Teaduste Akad. 4 : 265, 1983.

Fruitbody annual, reviving during the subsequent season, resupinate or effused-reflexed, imbricate and confluent, reflexed portion conchate to semi-circular, effused part easily separable from host, highly water-soaked during monsoon, brittle and light when dry; 4-6 × 3-4.5 × 1-3 cm. Pileus surface dull or dirty white to cream (4A3), older regions ash grey (1B2) to orange grey (5B2) near margin, azonate, uneven, hispid to scrupose, strigose towards margin. Pore surface white while fresh, cream (4A3) when dry; pores large and visible to unaided eye, angular, 5-6 sided, 0.5-1 per mm, smaller towards margin; dissepiments 120-150 µm. thick; pore tubes stratified, older layers compressed, concolorous with the pore surface, 0.3-0.6 cm long, oblique, sunk into equal depths. Context concolorous or darker than pore tubes, up to 1.5 cm thick, present as a thin layer (1 mm) in between pore tubes, homogeneous.

Hyphal system monomitic. Generative hyphae hyaline, slightly thick-walled with broad and distinct lumen, branched, simple-septate, 3-5 µm in diam.; context hyphae in the older region slightly brownish, encrusted and agglutinated, 3-5 um in diam. Basidia clavate to cylindric-clavate, 2 or 4-spored, sometimes with oil globules, 22.5-30 × 6.5-8 µm; sterigmata 2.5-4 µm long, straight. Cystidia hyaline, long, cylindrical, thin-to slightly thick-walled, sometimes septate, apically encrusted, up to 75 × 12.5 µm. Basidiospores hyaline, ovoid or broadly ellipsoid, slightly thick-walled, smooth, 0 to 1 guttulate, nonamyloid, (9) 11-13 × 7-8 µm; spore print white.

DECAY : White fibrous rot

SPECIMEN EXAMINED : On unidentified tree trunk, Kamathalamudi, Tunacadavu, Parambiculam, Palghat Dist., Herb. CALI No. G221, 12-7-1985.

GEOGRAPHICAL DISTRIBUTION : New Caledonia, Philippines, Sri Lanka, India, Zambia, Zaire, Australia.

REMARKS : Large, thick-walled spores and septate cystidia are characteristic of this species. The earlier Indian report (Bakshi, 1977) shows smaller spores and narrower cystidia. The Sri Lankan (Lloyd, 1904-1919) and African (Ryvarden & Johansen, 1980) sporophores also appear similar to the present collection but the latter has bigger cystidia and smaller spores. Parmasto (1983b) describes highly variable cystidia in Russian collections. Compared with the spores of different collection recorded by Parmasto (1983b), those of the present collection measure the largest.

PERENNIPORIA Murr.

Mycologia 34 : 595, 1942.

Fruitbodies mostly perennial, rarely annual, resupinate, effused-reflexed to pileate, dimidiate to broadly attached, consistency woody-hard to brittle, pileus slightly velvety to smooth, ochrceous, brown to blackish with age, pore surface white, cream to dirty straw-coloured, pores small, iso-diametric, tubes usually stratified, context thin, white to light ochraceous, hyphal system di-trimitic, generative hyphae thin-walled, hyaline, smooth and clamped, often difficult to observe, vegetative hyphae dominating in the fruitbody and often difficult to assign to one of the two groups of vegetative hyphae, solid to thick-walled, unbranched to moderately branched, non-dextrinoid to variably dextrinoid, cystidia absent or present, spores thick-walled, globose, oblong ellipsoids to pipshaped, often truncate, hyaline to yellow, variably dextinoid, often with spores reacting variably within the same fruitbody. On dead deciduous and coniferous wood. Cosmopolitan genus.

Type species : *Perenniporia medulla-panis* (Fr.) Donk.

PERENNIPORIA OCHROLEUCA (Berk.) Ryv.

Norw. J. Bot. 19 (3-4) : 233, 1972. Fig. 32. f-h; Pl., fig.

Polyporus ochroleucus Berk., Hook. Lond. J. Bot. 4 : 33, 1845

Trametes scrobiculata Berk., Grevillea 6 : 70, 1877

P. compressus Berk., Lond. J. Bot. 4 : 53, 1845

Fomitopsis ochroleuca (Berk.) Imaz., Tokyo Sci-Mus. 6 : 92, 1943

Fruitbody annual to perennial, solitary, sessile or attached with a narrow base 8mm to 2cm wide, 5mm to 1 cm thick, corky; Pileus orbicular, applanate to slightly ungulate, surface pale orange (5A3) to orange white (4A2-4A3) to greyish orange when dry (6B4-6B5) to almost light brown (7D4-7D5), sometimes concentrically zonate, not striate, sometimes ridgid, margin thick and round, entire. Pore surface whitish to cream (4A2-4A3) to greyish orange (6B4-6B5) to almost light brown when dry. Pores round, 2-5 per mm, dissepiments thick; pore tubes single or stratified creamy white (4A2) becoming very pale brown 5-8mm long, fertile almost up to margin. Context 1-3mm thick, orange white (5A2) not differentiated in any way.

Hyphal system trimetic, all hyphae hyaline, generative hyphae 3.1 to 5.3 µm broad thin-walled; skeletal hyphae thick-walled with distinct lumen, up to 4.5 µm broad, slightly tortuous; binding hyphae narrow, 2.5-3.5 µm diam; sparingly branched, weakly dextrinoid. Basidia not seen. Spores 12.5-15.5 × 8.5-9.5 (10) µm abundant, ellipsoid with a truncate end, doubled walled, hyaline, smooth differentially dextrinoid.

SPECIMEN EXAMINED : On dead twigs fallen on the ground and som

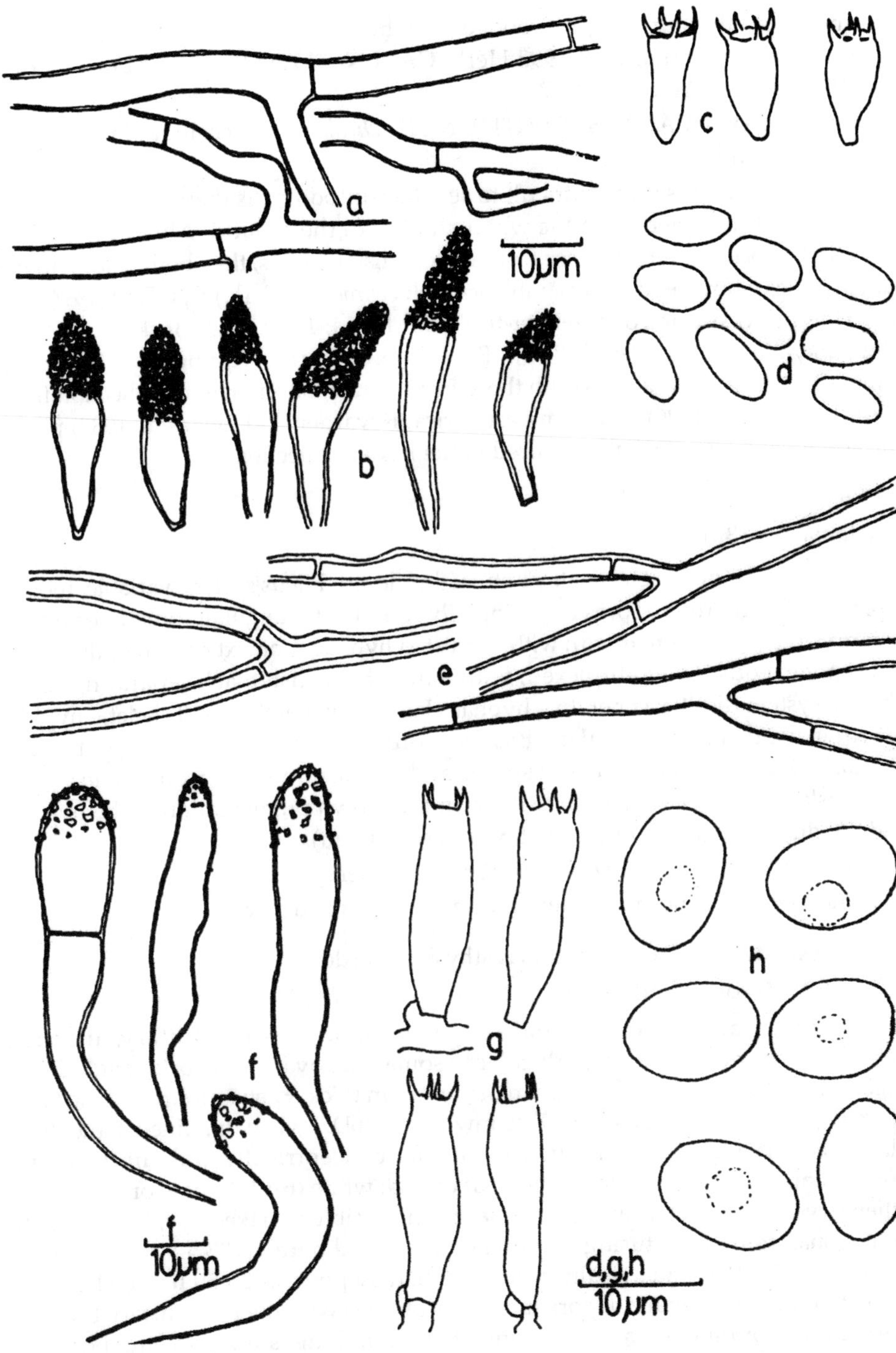

Fig. 23 : a-d *Oxyporus cervino-gilvus* : a-generative hyphae; b-cystidia; c-basidia; d-basidiospores.

e-h *Oxyporus mollissimus* : e-generative hyphae; f-cystidia; g-basidia; h-basidiospores.

detached from the substrate, in wooded area, Botanical garden, Calicut University Campus, Kerala, India, 13-8-1993 Herb. CALI No. F756; Kannur, Kerala, India, 10-8-1997, Herb. CALI. No. M567.

GEOGRAPHICAL DISTRIBUTION : *P. ochroleuca* is a cosmopolitan species reported from all the continents.

REMARKS : It is often difficult to see the fruitbodies as they are small in size and often get detached from the wood and fall to the ground. Although the two collections made from Kerala are from widely distant areas, they both belong to the same species, showing close resemblance to *P. ochroleuca* (Berk.) Ryv. Roy (1996) has described two species of *Perenneporia* from India, *P. martius* (Berk.) Ryv. from Meghalaya and *P. ochroleuca* (Berk.) Ryv. from Calcutta, West Bengal. The spore dimension of the latter agrees with those from Kerala, but unlike the collection from West Bengal (Roy, 1996) and from East Africa (Ryvarden and Johansen, 1980) pileus surface always remains pale coloured in the Kerala collections.

PILOPORIA Niemela

Karstenia 22 : 13, 1982.

Fruitbody annual, resupinate or effused-reflexed. Pileus surface variable, mostly sepia to reddish brown. Pores large, initially round, later angular, sometimes fusing to form large pores, pore mouth with encrusted hyphae. Context coloured, duplexed or homogeneous, when duplexed, with a thin, brownish black separating layer. Hyphal system dimitic; generative hyphae clamped, generative and skeletal hyphae brown in context and subiculam. Basidiospores allantoid or lunate. (Differs from *Datronia* Donk in having narrow spores and encrustohyphae, from *Incrustoporia* Domanski and *Skeletocutis* Kotl. & Pouz. in the brown pigmentation of the hyphae and the duplex structure of the context and subiculam).

Type species : *Piloporia sajanensis* (Parm.) Niemela

One species has been collected during the present survey.

PILOPORIA INDICA Ganesh, Leelavathy & Ryvarden

Trans. Brit. Mycol. Soc. 91(1) : 169-171. fig. 24. a-d; Pl., fig.

Fruitbody annual, pileate and semi-pendulous, sessile, solitary, attached broadly on the dorsal surface; orbicular to somewhat ovate, soft corky while fresh, slightly brittle when dry; 2.8-3.5 cm in diam; 0.8 cm thick. Pileus surface cognac (6E7) to agate (7E8), reddish golden (6C7) to raw Sienna (6D7) when dry; finely tomentose, glabrescent towards the margin, azonate but concentrically and faintly striate towards margin. Pore surface somewhat orange white (6A2) to pale orange (5A3) when dry, smooth but uneven; pores large and visible to naked eye, hexagonal to pentagonal, sometimes fusing to form large rounded pores, (1) 3.5-5 per mm; pore mouth 175-275 µm wide, pruinose to eroded; dissepiments 35-50 µm wide; pores not extending to the margin, pore tubes concolorous with the pore surface towards pore mouth, orange towards the context, 2-3 mm long, sunk into equal depths, straight, slightly friable. Context orange peel (6B8) while fresh and dry, 3-5 mm thick, spongy, more or less homogeneous, somewhat denser towards the tubes, without a distinct separating layer, cottony and tough.

Hyphal system dimitic. Generative hyphae hyaline, thin-to slightly thick-walled, branched, occasionally septate, clamped at each septa, 2.5-4 μm in diam. Skeletal hyphae orange to orange brown, darkening in KOH, less branched, thick-walled with narrow lumen, non-septate, 2.5-5 μm in diam. In dissepiments most hyphae have encrustations towards its extremities. Cystidia and cystidioles none. Basidia not observed. Basidiospores hyaline, allantoid to cylindric, thin-walled, smooth, nonamyloid; 3-4 × 0.3 0.7 μm.

DECAY : Associated with white fibrous rot on fallen twigs.

MATERIALS EXAMINED : On decaying twigs of *Prunus ceylanica* (Wight) Miq. (ROSACEAE), Kaduvakkunnu, Nadugani forests, Nilambur Forest Division, Malappuram Dist., Herb. CALI No. G140, 13-11-1984.

REMARKS : *Piloporia* is a monotypic genus erected by Niemela (1982) to accommodate *Antrodia sajanensis* Parm. which differs from other species of *Antrodia* Karst. in having a duplex context which is tobacco brown in colour while other species of *Antrodia* have a homogeneous and mostly white or light coloured context. In addition, species of *Antrodia* cause brown rot while *Piloporia* (=*A. Sajanensis*) cause white rot.

The present taxon which has orange brown context differs from *P. Sajanensis* (Parm.) Niemela in its colour and construction of the context. *P. Sajanensis* has a tobacco brown distinctly duplex context while in the present collection the context becomes slightly denser towards the tubes but without a distinct delimitation. In a way the present taxon is intermediate between *Piloporia* and *Antrodia* with more characters agreeing with *Piloporia* and hence it is tentatively kept in that genus.

The distinguishing features of this species are the semi-pendulous, sessile, solitary reddish brown fruibtodies, dimitic hyphal system with orange brown skeletal hyphae darkening in KOH, encrustation of hyphae in the dissepiments and small allantoid basidiospores.

POLYPORUS S. Str. Fr.

Syst. Mycol. 1 : 341, 1821.

Fruitbody annual, centrally or laterally stipitate. Pileus surface light to deep brown or almost purplish when old, smooth to scaly or finely tomentose at first, becoming glabrescent. Pore surface white to cream, pores entire, round to angular, small to large. Stipe light to deep brown or even blackish, glabrous to finely tomentose, smooth to longitudinally wrinkled. Context white. Hyphal system dimitic; generative hyphae hyaline, thin-walled, clamped; binding hyphae arboriform to dendriform with long, mostly dichotomously branched segments ending with thin, whip-like tips. Cystidia none. Basidiospores hyaline, cylindrical, straight to slightly bent, thin-walled, smooth and nonamyloid. On dead deciduous wood, rarely on conifers.

Type species : *Polyporus squamosus.* Fr.

Six species have been collected from Kerala.

KEY TO SPECIES

1.	Stipe central to eccentric	2
1′.	Stipe distinctly lateral	5
2.	Fruitbody always on the wood	3
2′.	Fruitbody on the ground attached to grass roots	*Polyporus meridionalis*
3.	Margin without cilia; stipe black	*P. hemicapnodes*
3′.	Margin ciliate; stipe light yellow to brown	4
4.	Pores 3-5 per mm; dissepiments 40-65 μm thick; spores oblong-elliptical, 7.5-8.5 × 3-3.7 μm	*P. tricholoma*
4′.	Pores 1-3 per mm, in radiating rows; dissepiments 160-200 μm thick; spores cylindric-ellipsoid, 5-8 × 2-3 μm	*P. arcularius*
5.	Stipe reduced; pores 7-8 per mm; Spores ovoid, 5-7 × 3.5 μm	*P. dictyopus*
5′.	Sub-stipitate, pores 3.5-5 per mm; spores oblong-ellipsoid, 6-9 × 2.5-3.5 μm	*P. grammocephalus*

POLYPORUS ARCULARIUS (Batsch) Fr.

Syst. Mycol. 1 : 342, 1821.

fig. 24. e-h.; Pl., fig.

Boletus arcularius Batch, Elench Fung. P. 97, 1783.

Polyporus collybioides Kalch. ex Cke., Grevillea 10 : 94, 1882.

Leucoporus arcularius (Batsch) Quel., Fl. Mycol. Fr. P. 402. 1888.

Polyporus penningtonii Speg., Mycol. Argent. 2 : 52, 1902.

P. arculariellus Murr., Bull. Torrey Bot. Club 31 : 36, 1904.

Favolus arcularius (Fr.) Ames, Ann. Mycol. 11 : 241, 1913.

Fruitbody annual, pileate, stipitate, either solitary or in small groups, sometimes more than one fruitbody arising from a common base, orbicular, central notched, coriaceous while fresh, brittle when dry; 1.5-5 cm in diam., 1-3 mm thick. Pileus surface yellowish brown (5C8) to light brown (5D8), brown (6E8) at the notch; tesselate, finely tomentose while young, glabrous when mature; margin thin and papery, densely ciliate, cilia up to 1 mm long, margin inrolled when dry. Stipe centraly or rarely eccentric, pale yellow (4A4) or slightly brownish, finely tomentose to glabrascent, cylindrical, bulbous at the base, solid; stipe-tissue cream (4A3); 1.2-3.5 cm long, 1.5-2.5 mm thick. Pore surface white with a greenish tinge, cream-coloured when dry; pores slightly decurrent, absent near margin, large and conspicuous, smaller towards the margin and stipe, angular, pentagonal or polygonal in radiating rows, 1-3 per mm; dissepiments torn at the pore mouth, 160-200 μm thick; pore tubes concolorous with the pore surface, sunk into equal depths, 1-2 mm long. Context white, homogeneous, up to 0.5 mm thick.

Hyphal system dimitic. Generative hyphae hyaline, thin-walled, branched, nodose-septate, often breaking at nodes, 2.5-5 μm in diam., frequent. Binding hyphae hyaline, thick-walled with narrow to obliterating lumen, solid towards ending, branched profusely, non-septate, attenuated towards extremities, hyphae often twisted, 2.5-5 μm in diam. Hyphal pegs rare. Basidia clavate, 2 or 4-spored, 13.5-18 × 5.25-6.75 μm; sterigmata erect, up to 4 μm long. Cystidia none. Basidiospores hyaline, thin-walled, cylindric to cylindric ellipsoid, smooth, 0 to 2-guttulate, nonamyloid (5)6-8 × 2.2-3 μm. Spore print white.

DECAY : White fibrous rot.

SPECIMEN EXAMINED : On decaying branches, host not identified, Botanical Garden, Calicut University Campus, Malappuram Dist., Herb. CALI No. G2, 17-6-1983; G166, 23-5-1985; Mundakai, Meppadi, Wyanaad Dist., Herb. CALI No. G115, G118, 22-9-1984; Karadimala, Brahmagiri Ranges, Tirunelli, Wyanaad Dist., Herb. CALI No. G257, 2-10-1985; Chooralmala, Meppadi, Wyanaad Dist., Herb. CALI No. G251, 13-9-1985; Vengoli Peak, Teak Plantation Division, Tunacadavu, Palghat Dist., Herb. CALI No. G265, 12-11-1985.

GEOGRAPHICAL DISTRIBUTION : Cosmopolitan.

REMARKS : The earlier Indian collections (Bose, 1991-1928; Thind & Chatrath, 1957; Bakshi, 1971) of this fungus agree with the present collection in macro- and microcharacters but for the ovate and much smaller basidiospores described by Bose (1919-1928). The Tropical American (Murrill, 1915), North American (Lowe, 1942) and New Zealandean (Cunningham, 1965) collections of *P. arcularius* also agree with the present collection while those from Western United States (Lowe & Gilbertson, 1961) and East Africa (Ryvarden & Johansen, 1980) have slightly longer spores.

POLYPORUS DICTYOPUS Mont.

Ann. Sci. Nat. 2, 3 : 349, 1835. fig. 24. i-l.; Pl. fig.

Melanopus dictyopus (Mont.) Pat., Essai Tax. P. 80, 1900.

Fruitbody annual, attached with a short lateral stipe, imbricate, confluent, dimidiate to flabelliform, applanate, sometimes incised into lobes, coriaceous while fresh, tough when dry; 4.5-7 × 3-4 × -0.35 cm. Pileus surface corn (4B5) with some patches of hair brown (5E4), sometimes greyish ruby (12E3), radially striate, glabrous, wrinkled when dry; margin wavy, entire to crenate, reflexed when dry. Stipe reduced, 2-3 mm long, 2-3 mm thick, slightly discoid at the base, solid; stipe-tissue cream

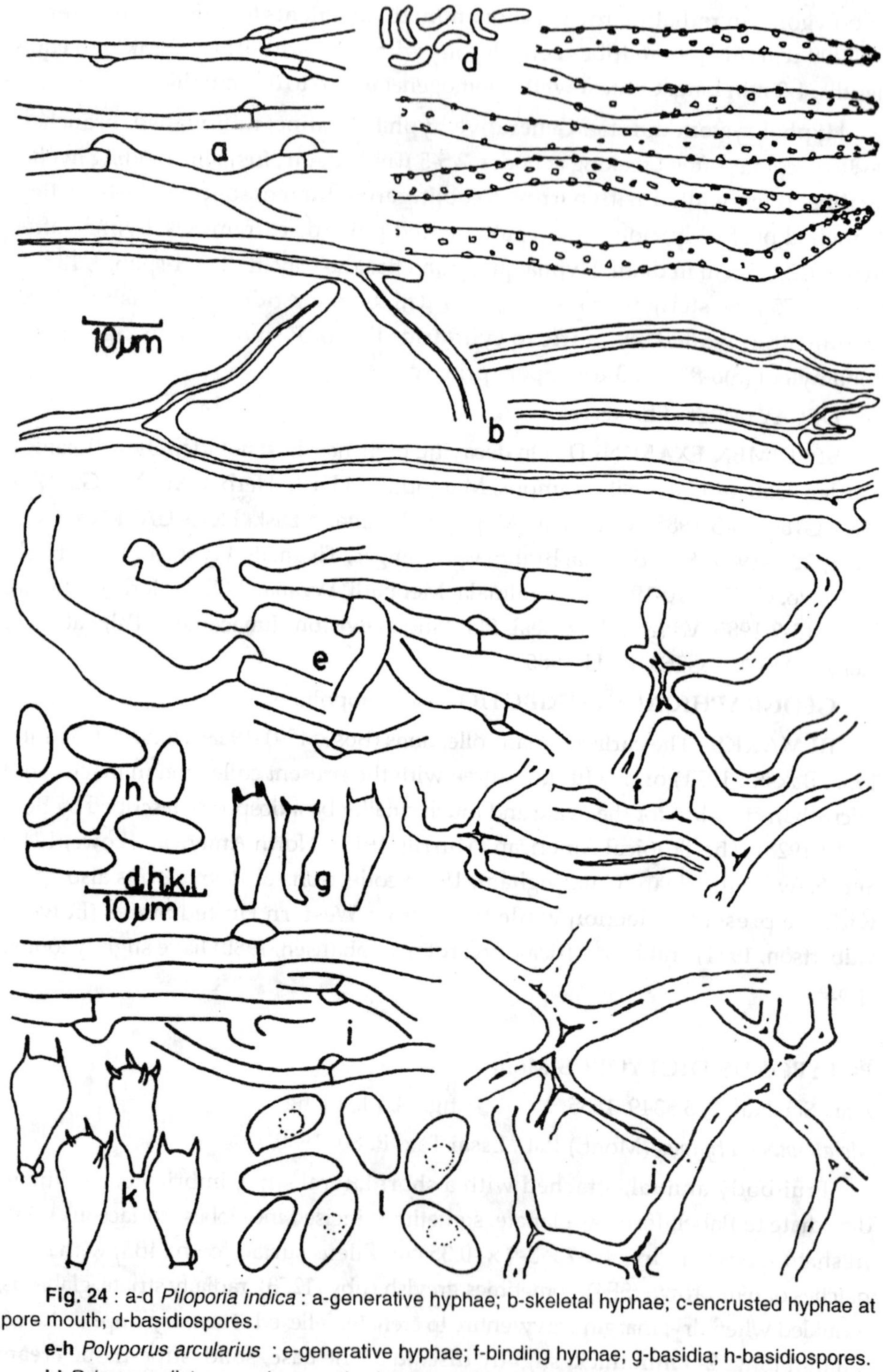

Fig. 24 : a-d *Piloporia indica* : a-generative hyphae; b-skeletal hyphae; c-encrusted hyphae at pore mouth; d-basidiospores.
e-h *Polyporus arcularius* : e-generative hyphae; f-binding hyphae; g-basidia; h-basidiospores.
i-l *Polyporus dictyopus* : i-generative hyphae; j-binding hyphae; k-basidia; l-basidiospores.

(4A3). Pore surface pale yellowish white (2A2) becoming yellowish white (3A2) on bruishing; pores minute and not visible to unaided eye, absent near margin, decurrent, round, regular, 7-8 per mm; pore mouth 75-100 μm wide; dissepiments 35-50 μm thick; pore tubes concolorous with the pore surface, up to 2 mm long, sunk into equal depths, straight. Context almost yellowish white (3A2), up to 1.5 mm thick.

Hyphal system dimitic. Generative hyphae hyaline, thin-walled, branched, sepatate with clamps, frequently present in context, easily breaking at nodes, 3-4.5 μm in diam. Binding hyphae pale-coloured, highly branched, thick-walled to solid with obliterate lumen, non-septate, flexuous, 2.5-3 μm in diam. Basidia clavate, 2 or 4 spored, 13-16.5 × 6-7.5 μm; sterigmata up to 3 μm long; basidioles 10-13 × 6-7.5 μm; frequent. Basidiospores hyaline, ovoid to ellipsoid, thin-walled, smooth, 0 to 2 guttulate usually uniguttulate, 5-7 × 3.5-4 μm.

DECAY : White rot.

SPECIMEN EXAMINED : On decaying branchlets lying on the forest floor, host not identified, Chembra Peak, Meppadi, Wyanaad Dist., Herb. CALI No. G235, 11.9.1985; on fallen branchlets, Mundakai, Meppadi, Wyanaad Dist., Herb. CALI No. G250, 12-9-1985.

GEOGRAPHICAL DISTRUBTION : South America, Australia, New Zealand, India, Sri Lanka, Africa.

REMARKS : In India this is the only record of *P. dictyopus* outside of Bengal. Bakshi's (1971) description of the species (after Bose, 1946) agrees somewhat closely with the present collection except for slightly thinner generative hyphae in the former. Wright & Deschamps (1972) reports bigger basidia and elliptical spores for the Argentenian collections while Ryvarden & Johansen (1980) reports longer stipe, smaller pores and elliptical spores in East African sporophorers of *P. dictyopus*.

POLYPORUS GRAMMOCEPHALUS Berk.

Lond. J. Bot. 1 : 148, 1842. Pl. 10, fig. 25. a-d.; Pl., fig.

Polyporus muelleri Kalch. ex Cke., Grevillea 10 : 97, 1882.

P. albellus Mass., Kew Bull. mis. inf. 1889. P. 171, 1889 nomen illegit., non Peck., 1878.

Leucoporus grammocephalus (Berk.) Pat., Essai Tax. P. 82, 1900.

Polystictus sebesici Lloyd, Mycol. Notes 71 : 1240, 1924.

Tyromyces grammocephalus (Berk.) G.H. Cunn., N.Z. Dep. Sci. Industr. Res. Bull. 164 : 135, 1965.

Fruitbody epixylous, annual, pileate, attached laterally with a sub-stipe, solitary to imbricate, dimidiate to flabelliform, applanate, sometimes lobate towards margin, coriaceous while fresh, becoming tough when dry; 5-14.5 X 4-10 X 0.2-0.5 cm. Pileus surface orange yellow (4A8), brownish orange (6C8), coganc (6E7), greyish orange

(5B5) or bronze (5E5); radially striate, striations more prominent towards base, glabrous, frequently with concentric wrinkles, sometimes context hyphae arise in tufts towards base; margin thin, entire, reflexed when dry. Stipe reduced or well developed, concolorous with pileus surface, solid, base expanding; stipe-tissue cream-coloured. Pore surface brownish yellow (5C8), to light orange (5A4); pores small and visible to naked eye, semi-decurrent, extending to the margin, round to angular, 4 to 6-sided when angular, 3.5-5 per mm; pore mouth 120-180 μm broad; dissepiments 60-90 μm thick; pore tubes concolorous with the pore surface, 1-1.5 mm long, straight, sunk into equal depths, shallow towards margin. Context cream (4A3) to butter yellow (4A5), homogeneous, 1-3 mm thick.

Hyphal system dimitic. Generative hyphae hyaline, thin-walled, sparsely nodose septate, common in dissepiments and younger areas, 1.5-2.5 μm in diam. Binding hyphae hyaline to pale-coloured, thick-walled, copiously branched, non-septate, cylindrical to slightly flexuous, 2.5-5 μm in diam. Cystidia none. Basidia broadly clavate, 2 or 4-spored, 15-17.5 × 6.25 μm; stergmata up to 3 μm long. Basidiospores hyaline, oblong-ellipsoid, thin-walled, smooth, nonamyloid, 6-9 × 2.5-3.5 μm.

DECAY : White stringly rot.

SPECIMEN EXAMINED : On decaying branches of *Macaranga peltata* (Roxb.) Muel.-Arg. (EUPHORBIACEAE), University Campus, Malappuram Dist., Herb. CALI No. G323, 6-7-1986, G359, 31-8-1986; on dead branches of *Artrocarpus heretophyllus* Lamk. (MORACEAE), Chakkumpuzha, Kottayan Dist., Herb. CALI No. G11, 30-6-1983; Calicut University Campus, Mallapuram Dist., Herb. CALI No. G184, 11-6-1985; on decaying stumps of *Laportea crenulata* Gaudich. (URTICEAE), Perumpara (W), Malakkapara, Sholayar forests, Trichur Dist., Herb. CALI No. G335, 29-8-1986; on fallen branches of *Callicarpa tomentosa* (Linn.) Murray (VERBENACEAE), Pringal, Sholar forests, Trichur Dist., Herb. CALI No. G8, 28-6-1983; on decaying forest logs, host unidentified, Nedumgayam, Nilambur, Malappuram Dist., Herb. CALI No. G24, 19-3-1983; Kanjirakadavu, Amarambalam Range, Nilambur, Malappuram Dist., Herb. CALI No. G106, 30-6-1985; Pathenthode, Silent Valley National Park, Palghat Dist., Herb. CALI No. G56, 30-5-1984; G108, 7-9-1983; Chooralmala, Meppadi, Wyanaad Dist., Herb. CALI No. G255, 13-9-1985; Noolpuzha, Muthanga Reserve Forests, Wyanaad Dist., Herb. CALI No. G313, 22-6-1986; Perumpara (W), Sholayar forests, Trichur Dist., Herb. CALI No. G331, 29-8-1986.

REMARKS : One of the common polypores, *Polyporus grammocephalus* has been frequently collected and described from India (Bose, 1919-1928, 1937a; Thind & Chatrath, 1957; Bakshi 1971). The collection of Bose (1937a) shows bigger spores than those of any other collection. The description of this polypore by Thind &

Chatrath (1957) shows bigger pores, thicker dissepiments and hyphae while the description of Bakshi (1971) shows wider hyphae than those of the present collections. The New Zealandean (Cunningham, 1965) fruitbodies exhibit wider pores and dissepiments and much smaller spores, while the East African (Ryvarden & Johansen, 1980) fruitbodies have smaller pores than those of the present collection. This very common popypore was available throughout the rainy seasons on varied hosts from virgin evergreen forest to inhabited area.

POLYPORUS HEMICAPNODES Berk. & Br.

Linn. Soc. 14 : 47, 1873. fig. 25. e-h.; Pl., fig.

Polyporus pusillus Rostr., Bot. Tidasskr. 24 : 359, 1902.

Fruitbody annual, centrally stipitate, solitary, orbicular with a notched centre, up to 3 cm in diam., coriaceous in consistency. Pileus surface Pompeian yellow (5C6) to clay (5D5) or light orange (5B4) , slightly darker near the notch; glabrous, radially ridged, ridges more prominent when dry; margin dry. Stipe black, dark brown (6F8) near the apex, fine adpressed tomentum, almost putty (4B2) in colour, visible with a lens, slightly expanded towards the base, cylindrical, solid, 1.5-5 cm long, 1.5-2 mm thick; stipe tissue white. Pore surface yellowish white (2A2) to pale yellow (3A3); pores never decurrent, not extending to the margin, angular to round, 6-7 per mm; pore mouth 70-100 μm broad; dissepiments 30-45 μm thick; pore tubes cream, up to 1 mm long, sunk into equal depths. Context cream (4A3), up to 1 mm thick.

Hyphal system dimitic. Generative hyphae hyaline, thin-walled, branched, septate with clamps, common in context and dissepiments, 2.5-3(5) μm in diam. Binding hyphae hyaline, thick-walled to solid, profusely branched, non-septate, slightly flexuous, tips attenuated, 2-2.5(7) μm in diam., common in context and stipe. Basidia clavate, 2 or 4- spored, 15-20 × 6.25-7.5 μm; sterigmata 2.5-2.75 μm long, cystidia absent. Basidiospores cylindrical, thin-walled, hyaline, 1 to 2-guttulate, (6)7-8.5(10) × 2.5-3.5 μm.

DECAY : White fibrous rot.

SPECIMEN EXAMINED : On decaying branchlets, host not identified, Perumpara (W), Sholayar forests, Trichur Dist., Herb. CALI No. G169, 10-6-1985.

GEOGRAPHICAL DISTRIBUTION : Sri Lanka, India, Thailand, Africa.

REMARKS : This is the first record of this not so common polypore from India. Originally described from Sri Lanka (Berkeley, 1872) it is apparently rare, usually small and easily overlooked according to Ryvarden (1976d) who made type studies of species described by Rostrup from Thailand. The East African collections (Ryvarden & Johansen, 1980) agree colsely with the present collection except for bigger sporocarps in the former. The sporocarps and spores of *P. hemicapnodes* from Thailand (Ryvarden, 1976d) are much smaller than those from India and East Africa.

POLYPORUS TRICHOLOMA Mont.

Ann. Sci. Nat. 2, 8 : 365, 1837. fig. 25. i-l.; Pl., fig.

Polyporus raphanipes Wakef. Kew Bull. Misc. Inf. 1914 p. 154, 1914.

Fruitbody annual, solitary or in small gregarious groups, centrally stipitate, slightly infundibuliform or notched at the centre, orbicular, soft leathery while fresh, rigid when dry; 2.5-5.2 cm in diam., 3-4 mm thick. Pileus surface yellowish grey (4B2), greyish orange (5B3) to brownish orange (5C3), sometimes Pompeian yellow (5C6); glabrous, azonate, radially striate, slightly shiny, reflexed when dry; margin thinner, densely ciliate. Stipe concolorous with the pileus surface to brownish towards base, cylindrical, solid, 1.7-3 cm long, 0.2-0.5 cm thick; stipe-tissue white. Pore surface yellowish white (3A2), smooth and even; pores small and almost visible to naked eye, never decurrent, absent near margin, slightly angular, regular, 3-5 per mm; pore mouth 150-250 µm wide; dissepiments 40-65 µm thick; pore tubes white, sunk into equal depths, straight, upto 2 mm long. Context white, homogeneous, up to 2 mm thick.

Hyphal system dimitic. Generative hyphae hyaline, thin-walled, branched, septate with clamps, usually 2-3 µm in diam; often swollen hyphae up to 10 µm in diam, common in context and dissepiments. Binding hyphae hyaline or pale-coloured, thick-walled with distinct lumen, branched, non-septate, extremities twisted and attenuated, 2.5--7.5 µm in diam; swollen hyphae up to 32 µm diam. common in dissepiments. Basidia clavate, 4-spored, 12.5-15.5 × 5-7.5 µm; sterigmata slightly curved, 3.5-5 µm long. Basidioles numberous, almost 12-15 × 5-7.5 µm. Basidiospores hyaline, oblong-ellipitcal, smooth, thin-walled, eguttulate, nonamyloid, 7.5-8.7 × 3-3.7 µm; Spore print hyaline.

DECAY : White rot.

SPECIMEN EXAMINED : On decaying logs of *Anacardium occidentale* Linn. (ANACARDIACEAE), Calicut University campus, Malappuram Dist., Herb. CALI No. G59, 11-6-1984, G167, 29-5-1985; on unidentified host, Calicut University campus, Malappuram Dist., Herb. CALI No. G286, 16-6-1986.

GEOGRAPHICAL DISTRIBUTION : Brazil, Cambodia, Cuba, India, Jamaica, Mexico, Nigeria, Puerto Rico, USA.

REMARKS : *P. tricholoma* regularly occur in the wooded localities in and around Calicut University Campus. In India this polypore has earlier been recorded only from West Bengal (Roy & De, 1977) and these have bigger basidia and spores. The present collection closely resembles those described by Lloyd (1912) and Overholts (1953). The tropical American collection (Murrill, 1915) has narrower spores.

POLYPORUS MERIDIONALIS (David) Jahn, Fig. 32. i-l Pl., Fig.

Westfal Pilzbriefe 11 : 176, 1980

Leucoporus Meridionalis David, Bull. Soc. Mycol. Fr. 88 : 301, 1972

Basidiomes on the ground, very small in size, centrally stipitate. Pileus 0.8-1.5 mm in diam., slightly infundifuliform to almost plane, about 1 mm thick, surface

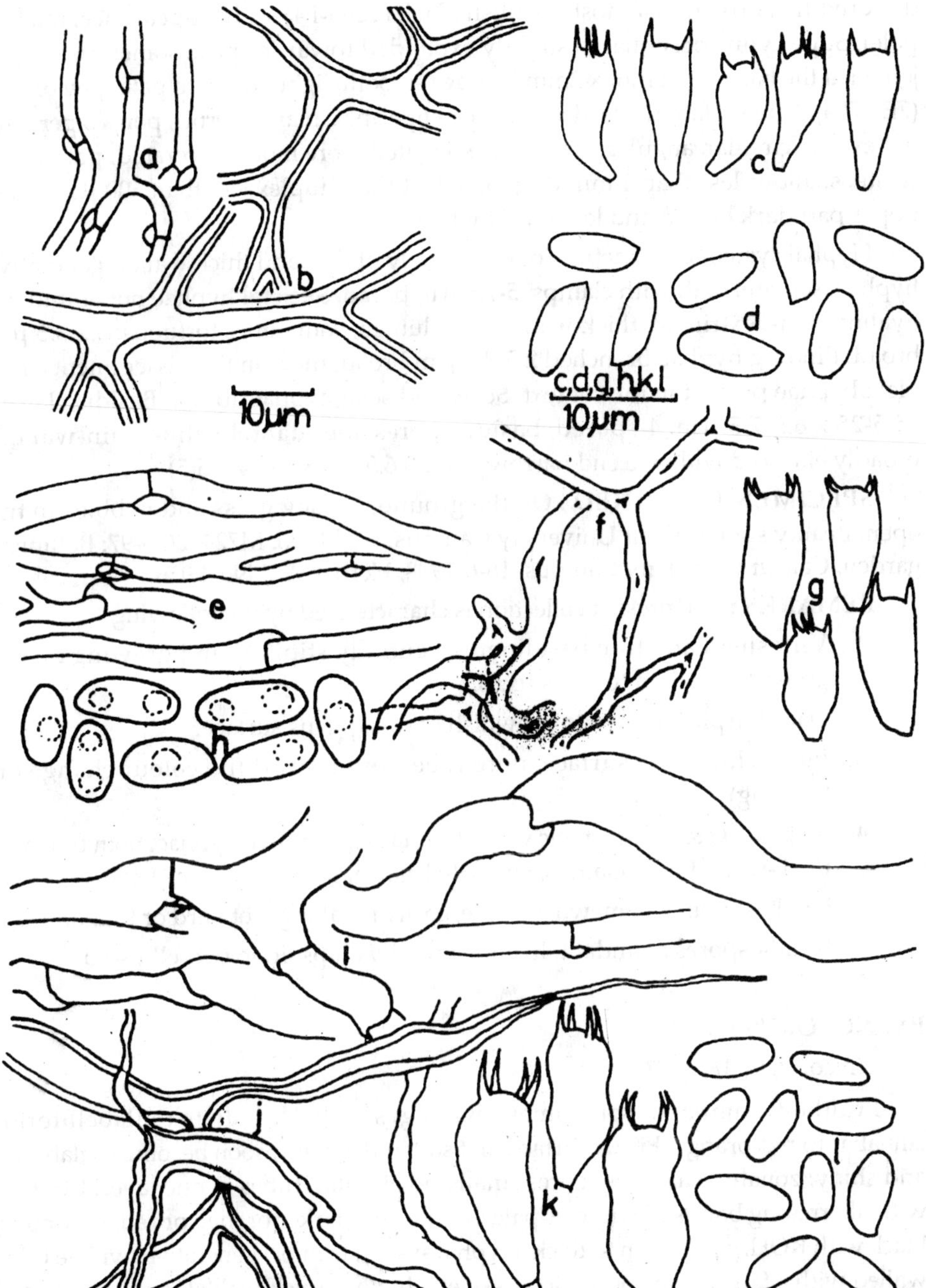

Fig. 25 : a-d *Polyporus grammocephalus* : a-generative hypahe; b-binding hypahe; c-basidia; d-basidiospores.

e-h *Polyporus hemicapnodes* : e-generative hypahe; f-binding hyphae; g-basidia; h-basidiospores.

i-l *Polyporus tricholoma* : i-generative hyphae; j-binding hyphae; k-basidia; l-basidiospores.

dark reddish brown to almost purplish black (12F3-F4) scaly appearance, margin paler. Stipe cylindric, equal to slightly expanded towards apex., sometimes stipes joined at the base, 11-16mm × 2mm, surface smooth, light brown to pale grey brown (7E3-7E4). Pore surface white, decurrent on stipe apex; margin fertile; pores 3 per mm, not exactly circular, angular to slightly elongated, pore mouth lacerated; pore tubes white, shallow, less than 1 mm deep. Context thin, duplex less than 500 μm thick, upper part dark brown and lower part white.

Hyphal system di-trimetic. Context mainly of thin and thick walled generative hyphae, septate, both with clamps, 5-7.5 μm broad. Skeletal hyphae not abundant, hyaline, non-dextrinoid, thick-walled with lumen, sometimes tortuous, 2.2-3.5 μm broad. Binding hyphae branched 2.5-3.5 μm broad, more in the dissepiments, few gloeohyphae present in the context. Setae and setal hyphae absent. Basidia clavate 15.5-25 × 6.5-7.5 μm, 4-spored; basidiospores abundant, hyaline, thin-walled, broadly ellipsoid with one end narrower, (6.2) 6.5-7.5(8.5) × 4-4.5(5).

SPECIMEN EXAMINED : On the ground, among grass and pebbles in the open, Beauty spot, Calicut University Campus, CALI No. M724, 26-8-97; Botanical garden, Calicut University Campus, 16-6-1998, Herb. CALI No. F956.

REMARKS : The present collection is characterised by the following :

1. Very small and thin basidiomes, centrally stipitate and growing on the ground/attached to grass roots.
2. Dark purplish brown to brownish orange pileus surface.
3. Pure white pore surface, pore tubes and context (no colour change on bruising)
4. Hyphal system di-trimetic with abundant generative hyphae, both thin and thick-walled with clamps, sparsely branched.
5. Dry basidiomes somewhat brittle, easily breaking, not hard or tough.
6. Basidiospores abundant, hyaline broadly ellipsoid to ovo-ellipsoid.

PYCNOPORUS Karst.

Rev. Mycol. 3, 9 : 18, 1881.

Fruitbody annual or sometimes reviving, sessile, dimidiate to flabelliform, cinnabar to red-orange. Pileus surface first soft and velvety, soon becoming glabrous and shiny, zonate to azonate. Pores small and circular, tubes in one layer. Context with alternating bands of white or pale orange and darker orange or red, becoming black with KOH, corky, thin to thick. Hyphal system trimitic; generative hyphae thin-walled with a few clamp-connections; skeletal hyphae thick-walled to solid, covered with orange crystals, prevailing in all parts of the fruitbody; binding hyphae thick-walled to solid, intensively tortuous and branched, with short lateral branches, numerous in the context just above the tubes. Cystidia absent. Basidia clavate with four sterigmata. Basidiospores short-cylindric, hyaline, nonamyloid and with smooth walls, about 4-6 μm long.

Type species : *Pycnoporus cinnabarinus* (Jacq. ex Fr.) Karst.
[=*P. sanguineus* (Linn. ex Fr.) Murr.]

One species has been collected from Kerala.

PYCNOPORUS SANGUINEUS (Linn. ex. Fr.) Murr.
Bull. Torrey Bot. Club 31 : 421, 1904, fig. 26. a-e; Pl. fig.

Boletus sanguineus Linn. Sp. Plant. P. 1646, 1763.

Polyporus sanguineus Linn. ex. Fr., Syst. Mycol. 1 : 371, 1821.

B. cinnabarina Jacq., Fl. Austr. 4 : 2, 1776.

Polyporus cinnabarinus (Jacq. ex Fr.) Fr., Syst. Mycol. 1 : 371, 1821.

P. coccineus Fr., Nova Acta 3, 1 : 67, 1851.

P. puniceus Kalch., Rev. Mycol. 4 : 95, 1882.

Polystictus sanguineus (Fr.) Cke., Grevillea 14 : 79, 1886.

P. cinnabarinus (Jacq. ex Fr.) Cke., Ibid 14 : 82, 1886.

Phellinus cinnabarinus (Jacq. ex Fr.) Quel., Fl. Mycol. Fr. P. 395, 1888.

Pycnoporus cinnabarinus (Jacq. ex Fr.) Karst., Rev. Mycol. 3 : 18, 1891.

Microporus sanguineus Pat., Essai tax. P. 83, 1900.

Polystictus semisanguineus Lloyd, Mycol. Letter 39 : 8, 1912.

Coriolus cinnabarinus (Jacq. ex Fr.) G.H. Cunn., N.Z. Dep. Sci. Industr. Res. Pl. Dis. Bull. 75 : 8, 1948.

C. sanguineus (Fr.) G.H. Cunn., Ibid. 81 : 17, 1949.

Fruitbody annual, pileate, sessile, arising in small gregarious groups, often solitary, imbricate and confluent, attached lateraly either with a broad base or with a converged narrow base; dimidiate to flabelliform, sometimes marginally lobed, lobes overlaping, coriaceous and flexible while fresh, slightly tough when dry; 2.5-7 × 2.4 × 0.1-0.4 cm. Pileus surface with concentric striations of reddish orange (7A7, 7B7), red-hair (6C4) or light orange (5A4), young sporophores totally white in some cases; glabrous, slightly crusty towards the base, radially wrinkled when dry; margin thin, edges eroded, rigid when dry. Pore surface reddish orange (7Ab) to tomato red (7C8), slightly rough; pores small but almost visible to unaided eye, margin sterile and distinct, round to oval, 3-4(5) per mm; pore mouth 140-180(200) µm wide; dissepiments 25-50 µm thick; pileus surface also develop pores and fertile hymenium in some sporophores which were upside down; pore tubes concolorous with the pore surface, straight and sunk into equal depths, 1-2 mm long. Context greyish red (7B6) to reddish orange (7B7), darker in KOH, up to 2 mm thick.

Hyphal system trimitic. Generative hyphae hyaline, thin-walled, septate with clamps, short-celled, branched, 1-5-4 µm in diam. Binding hyphae reddish yellow, thick-walled with narrow but distinct lumen, branched, non-septate, slightly attenuated towards extremities, 2.5-5 µm in diam. Skeletal hyphae reddish yellow, thick-walled with distinct lumen, unbranched or with short terminal branches, nonseptate, main stem cylindrical and equal in width, 3.75-7.5 μm in diam. Basidia clavate, 4-spored, 10-12.75 3-3.75 μm; sterigmata up to 3 μm long, straight. Basidiospores hyaline cylindrical to slightly ovate, smooth, thin-walled, non-amyloid, 3.5-4.5 × 1.75-1.5 μm; spore print white.

DECAY : White stringy rot, reddish mycelial sheets prominent in wood crevices.

SPECIMEN EXAMINED : On decaying logs of *Terminalia bellarica* (Gaertn) Roxb. (COMBRETACEAE), Kuppadi, Sultan Battery, Wyanad Dist., Herb. CALI No. G287, 20-6-1986; on decaying trunks of *Cocos nucifera* Linn. (PALMAE), Padikkal, near Calicut University campus, Malappuram Dist., Herb. CALI No. G87, 7-8-1984; Botanical Garden, Calicut University campus, Malappuram Dist., Herb. CALI No. G380, 29-9-1987; on decaying branchlets, host not identified, Silent Valley National Park, Palghat Dist., Herb. CALI No. G42, 15-1-1984.

GEOGRAPHICAL DISTRIBUTION : Cosmopolitan.

REMARKS : *P. sanguineus* is one of the most colourful polypores in this region. The crimson-coloured sporophores with large pores, association with white rot and ramifying red mycelium in the decaying wood are most distinguishing of *P. sanguineus*. Hyphal pegs reported in North Amercian sporophores (Lowe, 1934; Overholts, 1953) sems to be absent in tropical specimens including the present collections. The bleached sporophores reported by earlier workers (Lloyd, 1912; Bose, 1934) were found among the present collection also.

PYROFOMES Kotl. & Pouz.
Feddes Rep. 69 : 140, 1964.

Fruitbody annual to perennial, pileate to resupinate, broadly attached. Pileus surface smooth or tomentose, cream to reddish. Pore surface reddish brown to orange, pores small and entire, tubes reddish-brown to deep red. Context thin to conspicuous, reddish-brown to brick or deep reddish orange. Hyphal system dimitic or trimitic; generative hyphae thin-walled and with clamps; skeletal hyphae thick-walled to solid, straight to flexuous, unbranched to sparingly branched; binding hypahe sometimes in pore tubes. Cystidia and setae usually none. Basidiospores hyaline to coloured, thick-walled, truncate, nonamyloid. Decay white rot.

Type species : *Polyporus demidoffii* Lev. [=*Pyrofomes demidoffii* (Lev.) Kotl. & Pouz.]

Only one species has been recorded from Kerala.

PYROFOMES ALBOMARGINATUS (Zipp. ex Lev.) Ryv.
Norw. J. Bot. 19 : 236, 1972. fig. 26. f-g.; Pl., fig.

Polyporus albomarginatus Zipp. ex Lev., Ann. Sci. Nat. 3, 2 : 191, 1844.

P. kermes Berk. & Br., J. Linn. Soc. 14 : 49. 1873.

P. laeticolor Berk., Ibid. 16 : 46, 1877.

Fomes pyrrhoceras Cke., Grevillea 14 : 11, 1885.

F. kermes (Berk. & Br.) Cke., Ibid. 14 : 19, 1885.

F. albomarginatus (Zipp. ex Lev.) Cke., Ibid. 14 : 19, 1885.

Polyporus purpureoauranticus Beeli, Bull. Sco. Bot. Belg. 62 : 64, 1929.

Fomitopsis albomarginatus (Zipp. ex Lev.) Imaz., Bull. Govt. For. Exp. Sta. Tokyo 57
: 110, 1952.

Phellinus albomarginatus (Zipp. ex Lev.) G.H. Cunn., N.Z. Dep. Sci. Industr. Res. Bull. 164 : 239, 1965.

Fruitbody annual to perennial, pileate, sessile, effused-reflexed initially, later forming pileate fruitbodies, often imbricate and confluent, seldom solitary; broadly semi-circular to somewhat irregular, applanate, corky in consistency while fresh, hard when dry; 6.5-20 × 3.5-10 × 0.3-1 cm. Pileus surface dark brown (6F7) to coganc (6E7) towards the hase, margin usually eye brown (7E7) but sometimes brownish orange (6C8) or lighter; concentrically crusty and zonate, finely velutinate (at least towards margin), sometimes glabrous when matüre, underneath context appear scrupulose towards the periphery, rough and uneven; margin thin, rigid when dry. Pore surface brownish orange (6C8) to orange peel (6B8) while fresh becoming yellowish brown (5E8) with a slight orange tinge or copper red (7C8); smooth and wavy; pores minute and not visible to naked eye, absent near margin, round, regular, 8-9 per mm; pore mouth 75-85 µm wide; dissepiments 20-45 µm thick; pore tubes orange peel (6B8) while fresh, older layers copper (7C8) and older tubes inside whitish, stratified, 1-2 mm long in each stratum. Context concolorous with pore tubes, cherry red in KOH, floccose, each stratum 2-3 mm thick.

Hyphal system dimitic. Generative hyphae hyaline, thin-walled, sparsely branched, rarely septate, but clamped; difficult to observe in old specimens, 2.5-3.5 µm in diam. Skeletal hyphae reddish brown, almost cherry red in KOH, thick-walled with mostly broad lumen, often branched, non-septate, 4-6 µm in diam.; old tubes filled with binding hyphae, well branched, branchings short, non-septate, 2-3.5 µm in diam. Setae and cystidia absent. Basidia and basidiospores not observed.

DECAY : Associated with white rot. The fruitbody appear at the buttress base of an almost decayed tree. The region just below the sporophore was completely decayed into a white brittle mass.

SPECIMEN EXAMINED : On decaying tree base, host not identified, near Perumpara (W), Sholayar forests, Trichur Dist., Herb. CALI No. G345, 30-8-1986.

GEOGRAPHICAL DISTRIBUTION : West Indies, New Guinea, British Honduras, Africa, Australia, India, Bangaladesh, Sri Lanka, Thailand, Philippines, Malaysia, Sarawak, Indonesia.

REMARKS : Bakshi (1971) described it as *Fomes albomarginatus* with hyphal pegs and bigger pores. Murrill's (1915) description of British Honduras collection also have bigger pores and white margin. New Zealandean (Cunningham, 1965) and African (Ryvarden & Johansen, 1980) collections of *P. albomarginatus* are very similar to the present collection.

P. albomarginatus is characterised by non-rimose but soft zonate pileus surface, brick-red context, minute pores, clamped generative hyphae and mostly sterile hymenium.

RIGIDOPORUS Murr.

Bull. Torrey Bot. Club 32 : 478, 1905.

Fruitbody annual to perennial, resupinate to pileate. Pileus surface white to ochraceous to bright orange or pinkish in some tropical representatives, usually smooth to finely tomentose in narrow zones. Pore surface white, pink or reddish to light brown when dry, pores small and isodiametric. Context white to ochraceous, consistency soft to woody hard. Hyphal system monomitic, generative hypahe thin- to thick-walled and sparsely simple-septate. Cystidia absent or present, large, enclosed or projecting, thick-walled cystidia encrusted in the upper part; subulate to bottle shaped cystidioles present but easily deteriorating. Basidiospores smooth, thin- to thick-walled hyaline and nonamyloid. Mostly on deciduous wood.

Type species : *Polyporus micromegas* Mont. [=*Rigidoporus microporus* (Fr.) Overeem]

Four species have been collected from Kerala.

KEY TO SPECIES

1.	Encrusted cystidia present	2
1′.	Cystidia usually absent, if present mucronate, not encrusted	3
2.	Fruitbody pileate; cystidia 7.5-11 µm broad	*R. lineatus*
2′.	Fruitbody resupinate; cystidia 11-16.5 µm broad	*R. vinctus*
3.	Annual species; spores 4-5 µm or 4-5 × 3.5-4.5 um; decay causing white rot	*R. microporus*
3′.	Perennial species; spores 5-6.6 µm; decay causing brown cuboidal rot	*R. ulmarius*

RIGIDOPORUS LINEATUS (Bers.) Ryv.

Norw. J. Bot. 19 : 236, 1972. fig. 26. h-j.; Pl. fig.

Polyporus lineatus Pers. In Gaudichaud, voy. Aut. Monde, Paris p. 174, 1827.

P. zonalis Berk., Ann. Mag. Nat. Hist. 10 : 375, 1843.

Polyporus inconspicous Miq., Bull. Sci. Phys. Nat. Neerl. p. 454, 1839.

P. surinamensis Miq., Ibid. P. 454, 1839.

P. micromegas Mont., Ann. Sci. Nat. 2, 17 : 128, 1842.

P. plumbeus Lev., Ibid. 3, 5 : 136, 1846.

P. pusiolus Ces., Atti. Acad. Sci. Fisc. Nat. Napoli 8 : 5, 1879.

P. memorandus Speg., An. Soc. Cient Argent. 9 : 165, 1880.

Leptoporus zonalis (Berk.) Par., Essai Tax. P. 85, 1900.

Rigidioporus surinamensis (Miq.) Murr., Bull. Torrey Bot. Club. 34 : 473, 1907.

Polyporus yoshinagai Lloyd, Mycol. Letter 54 : 5, 1915.

Polystictus eburenus Lloyd, Mycol. Notes 73 : 1331, 1924.

Rigidioporus zonalis (Berk.) Imaz., Bull. Govt. For. Exp. Sta. 57 : 119, 1952.

Trichaptum zonale (Berk.) G.H. Cunn., N.Z. Dep. Sci. Industr. Res. Bull 165 : 100, 1965.

Fruitbody epixylous, annual, sometimes resupinate latter becoming pileate, attached with a slightly converging lateral base, sessile to sub-stiptate, dimidiate to flabelliform to spathulate, applanate, corky while fresh, woody when dry; 4-6.5 × 1.8-5 × 0.3-0.8 cm. Pileus surface light yellow (4A4), greyish orange (5B4), sometimes agate (7A8) to dark brown (7F8), concentrically zonate, slightly sulcate, radially wrinkled, finely velutinate while young, partly preserved or becoming glabrescent when mature; margin thin, sharp, decurving strongly on drying. Pore surface greyish orange (5B5) while young, becoming camel (6D4) when old, smooth and even; pores almost extending to the margin, never decurrent, very small, round to angular, regular, 7-9 per mm; pore mouth 85-100 µm wide; dissepiments 35-50 µm thick; margin dull white and distinct; pore tubes concolorous with the pore surface, sunk into equal depths, straight, up to 2 mm long. Context yellowish white (4A2) to orange white (5A2), up to 4 mm thick, thinning towards margin, homogeneous.

Hyphal system monomitic. Generative hyphae hyaline to cream-coloured, thin or thick-walled, usually unbranched or with short terminal branches, simple-septate, 2.5-7.5 µm in diam. Cystidia narrowly clavate, encrusted or capitate encrusted, thick-walled, 25-42.5 × 7.5-11 µm, abundant towards pore mouth. Basidia not observed. Basidiospores globose to sub-globose, hyaline, thin-walled, smooth, nonmayloid, 4.5-5.5 × 4.5 µm, with a large refractive oil globule.

DECAY : Associated with white pocket rot.

SPECIMEN EXAMINED : On decaying stumps of *Areca catechu* Linn. (PALMAE), Alpara, Peechi, Trichur Dist., Herb. CALI No. G16, 12-7-1983; on exposed roots of *Delonix regia* (Hook.) Rafin. (CAESALPINIAE), Alpara, Peechi, Trichur Dist., Herb. CALI No. G17, 12-7-1983; on logs of *Tectona grandis* Linn. f. (VERBENACEAE), Nedumgayam, Nilambur Forest Division, Malappuram Dist., Herb. CALI No. G22, 19-7-1983; Chetalayam, Wyanaad Dist., Herb. CALI No. G198, 18-6-1985; Kanjirakadavu, Nilambur Forest Division, Malappuram, Dist., Herb. CALI NO. G277, 1983; on stumps of *Calycopteris floribunda* (Roxb.) Poir (COMBRETACEAE), Pokkunnu, Calicut Dist., Herb. CALI No. G62, 21-6-1984; on decaying trunk of *Atrocarpus heterophyllus* Lam. (MORACEAE), Noolpuzha, Muthanga Reserve Forests, Wyanaad Dist., Herb. CALI No. G314, 22-6-1986; on decaying logs of *Hevea braziliensis* (H.B.&K.) Muell Arg. (EUPHORBIACEAE), Palappilly Estate, Trichur Dist., Herb. CALI No. G325, 27-6-1986; on logs of *Mallotus tetracoccus* (Roxb.) Kurz.

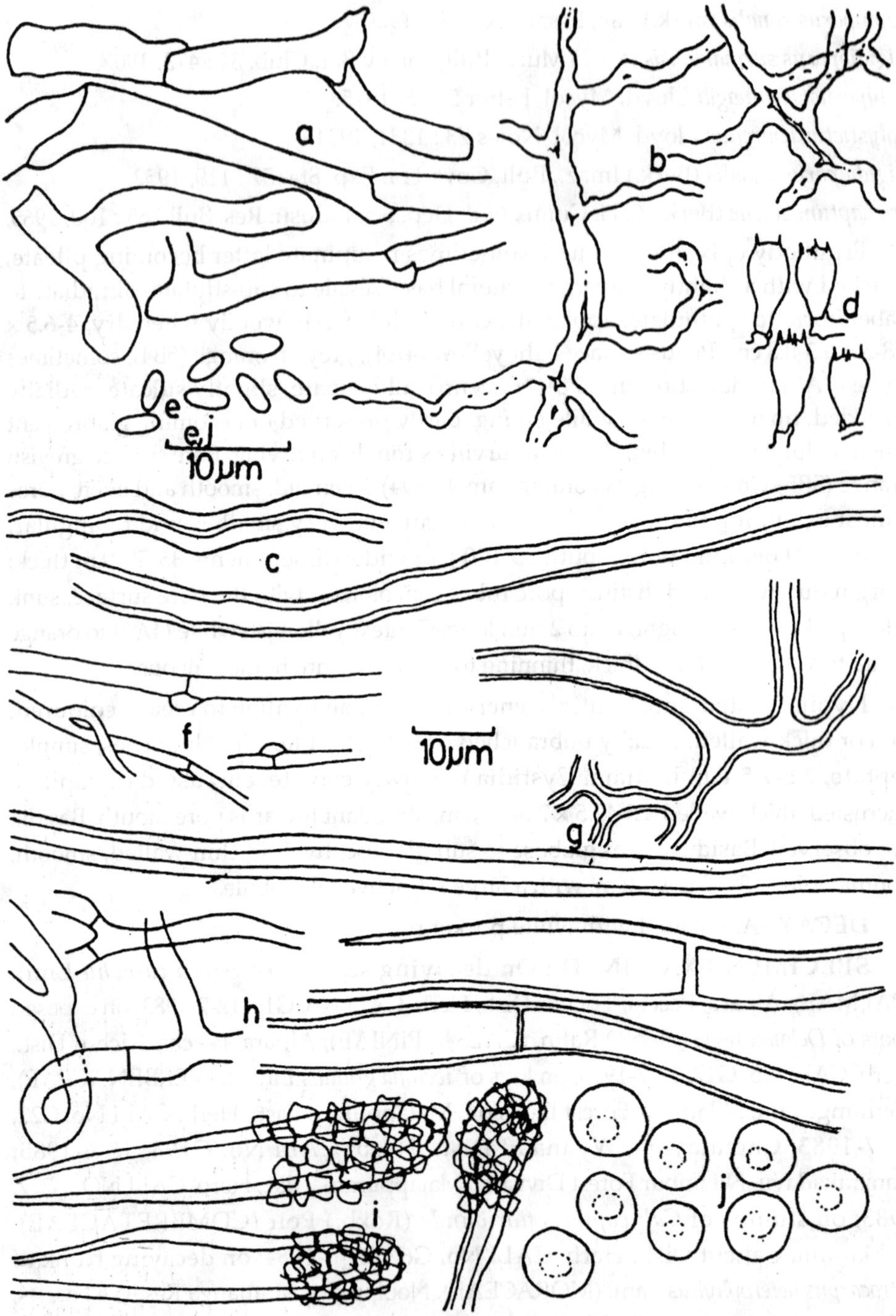

Fig. 26 : a-c *Pycnoporus sanguineus* : a-generative hyphae; b-binding hyphae; c-skeletal hyphae; d-basidia; e-basidiospores.

f-g *Pyrofomes albo-marginatus* : f-generative hyphae; g-skeletal hyphae;

h-j *Rigidoporus lineatus* : h-generative hyphae; i-cystidia; j-basidiospores.

(EUPHORIACEAE), Perumpara (W), Malakkapara, Sholayar forest Trichur Dist., Herb. CALI No. G355, 30-8-1986; on unidentified forest logs, Panthenthode, Silent Valley National Park, Palghat Dist., Herb. CALI No. G97, 7-9-1983; Kuppadi, Sultan Battery, Wyanaad Dist., Herb. CALI No. G295, 20-6-1986.

GEOGRAPHICAL DISTRIBUTION : India, Sri Lanka, North and South Africa, New Zealand, Australia, Fiji, Java, Malaya, Samoa, Tahiti, New Guinea, Mexico, Brazil, U.S.A.

REMARKS : This fungus has earlier been reported from India by Bose (1938), Bakshi et al. (1963) and Bakshi (1971) as *Polyporus zonalis*. Bakshi (1971) reports thin- and thick-walled generative hyphae and smaller spores. The description of North American collections by Overholts (1953) agrees well with the present material while Ryvarden & Johansen (1980) reports smaller cystidia and Pegler & Waterston (1968) reports freely branched generative hyphae. Strands of mycelium is seen growing on the wood just beneath the fruitbody the very next day after collection.

Occurring in natural forests, teak plantations and inhabited areas, the fruitbodies of *R. lineatus* were collected both on dead wood and living trees. In all cases the fruitbodies were growing near the base of the tree trunks or on exposed roots. In teak plantations, stumps remaining after the removal of poles during thinning operation, were mostly attached by *R. lineatus*. The fruitbodies start appearing from June, however, fertile collections were made during September. At one instance it was collected from the base of living *Calycopteris floribunda*, a woody climber. This species seems to be quite common and widely distributed in Kerala.

REGIDOPORUS MICROPORUS (Fr.) Overeem

Icon. Fung. Malayensum 5 : 1, 1924. fig. 27. a-b., Pl. , fig.

Polyporus microporus Fr., Syst. Mycol. 1 : 376, 1821.

P. lignosus Kl., Linnaea 8 : 485, 1833.

P. auberianus Mont., Ann. Sci. Nat. 2, 17 : 127, 1842.

P. cordubensis Speg., Ann. Mus. Nac. B. Aires 8 : 168, 1898.

Fomes lignosus (Kl.) Bres., Ann. Mycol. 10 : 496, 1912.

F. auberianus (Mont.) Murr., Bull. Torrey Bot. Club 32 : 491, 1905.

Fruitbody annual, seldom perennial, pileate, sessile, attached with a converging narrow or broad lateral base, sometimes effused-reflexed, imbricate and confluent, semi-circular to dimidiate or flabelliform, applanate, leathery to corky while fresh, rigid and light in weight when dry; 4-22 × 3-9 × 0.3-1.5 cm. Pileus surface light orange (5A5) to brownish orange (6C8) while actively growing, reddish brown (7E8) to yellowish brown (5E8) when dry, margin yellow (3A6) to butter yellow (4A5); concentrically striate to slightly furrowed, striations more prominent towards periphery, with shallow radial wrinkles, minutely pubescent while young, glabrous when old, often slightly crusty towards the base, fruitbody often incurved when dry; margin entire, thin, rigid. Pore surface wine yellow (3B3) while fresh, oyster grey (2C2) when mature, greyish orange (6B5) in certain specimens, smooth and even; pores minute and not visible to unaided eye, extending to the margin, round and regular 7-9 per mm; pore mouth 62-80 µm wide; dissepiments 20-35(50) µm thick; pore tubes concolorous with the pore surface or yellowish grey (3B2), 1-6mm long. Context cream (4A3), homogenous, up to 3 mm thick.

Hyphal system monomitic. Generative hyphae yellowish to light brown, thin to thick-walled with broad lumen, branched towards extremities, simple-septate, usually 3.75-7.5 µm; main stem broder, cylindrical and straight, 7.5-10 µm in diam. Cystidia absent Basidia not observed. Basidiospores hyaline, sub-globose to globose, thin-walled, smooth, nonaymyloid, eguttulate, 4-5 × 3. 5-4-5 µm.

DECAY : Associated with white rot.

SPECIMEN EXAMINED : On the trunk of living *Stereospermum colais* (Buch., Ham. ex. Dillw.) Mabberley (BIGNONIACEAE), Perumpara (W), Sholayar Forests Trichur Dist., Herb. CALI No. G171, 10-6-1985; on unknown forest logs, Begur range, Wyanaad Dist., Herb. FRI No. 4243, K. Bagchee, 6-4-1946.

GEOGRAPHICAL DISTRIBUTION : Africa, Australia, North, South and Central America, West Indies, India.

REMARKS : The earlier Indian report of this polypore (Bakshi et al, 1963; Bakshi 1971) shows mucronate cystidia. The CMI description of this fungs (Pegler & Waterston, 1968) shows larger pore mouth and smaller spores.

RIGIDOPORUS ULMARIUS (Sow. ex Fr.) Imaz.

Bull. Govt. For. Exp. Sta., Meguro 57 : 119, 1952.

fig. 27. g-h. Pl. , fig.

Boletus ulmarius Sow. Coloured Figures English Fungi, tab.P. 88; 1797.

Polyporus ulmarius (Sow.) ex Fr., Syst. Mycol. 1 : 365, 1821.

Fomes ulmarius (Sow. ex Fr.) Gillet, Champ. P. 683, 1878.

Polyporus xylocreon Speg., An. Soc. Cient. Argent. 16 : 128, 1880.

Fomes geotropus Cke., Grevillea 13 : 32, 1884.

Fomitopsis ulmatia (Sow. ex Fr.) Bond. & Sing, Ann. Mycol. 39 : 55, 1941.

Fruitbody usually perennial, solitary, effused-reflexed to pileate, sessile, attached with a broad lateral base, semi-circular, broadly applanate to conchate, usually with a drooping bell-like appearance, hard woody while fresh, tough and light when dry; 10-24 × 7.5-12 × 24 cm. Pileus surface pale yellow (4A3) to darker; slightly tomentose while young soon becoming glabrous, uneven and rough, concentrically zonate to deeply striate or even furrowed, crusty towards the base; margin thick, rounded and smooth. Pore surface Indian yellow (5B7) to yellow ochre (5C7), yellow

(5C7) to reddish blonde (5C4) to pale yellow (4A3) when dry, smooth and even; pores small and not visible to naked eye, absent near margin, round to oval, almost regular, 6-8 per mm; pore mouth 90-105 μm wide; dissepiments 40-50 (100)μm thick; thickness smoetimes varying, pore tubes Indian yellow (5B7) to apricot (5B6), almost most yellow (4A3) when dry, stratified, 4-15 mm long. Context pale yellow (4A3), hard corky, fibrous, 1.5-2.5 mm thick.

Hyphal system monomitic. Generative hyphae hyaline, thin to thick-walled, simple-septate, central stem sometimes branched, straight and equal in width, extremities freely branching, flexuous, attenuated, 2-5 μm in diam. Cystidia and basidia not observed. Basidiospores globose to subglobose, hyaline thin-walled, smooth, apiculate, with a large refractive guttule, nonamyloid, usually 5-6.5 μm in diam. Pileal cuticle a palisade of slightly clavate thick-walled hyphae.

DECAY : Causing brown cuboidal rot.

SPECIMEN EXAMINED : On the trunk of living *Schleichera oleosa* (Lour.) Oken (SAPINDACEAE), North Pathiri Reserve Forests, Perikaloor, Wyanaad Dist., Herb. CALI No. G307, 21-6-1986; on unidentified forest logs, Malakkapara, Sholayar forests, Trichur Dist., Herb. CALI No. G185, 10-6-1985.

GEOGRAPHICAL DISTRIBUTION : India, Japan, USSR, Cyprus, France, Germany, UK, Italy, Portugal, Mexico, British Honduras.

REMARKS : There are two earlier reports of this fungus from India. Bakshi (1971) recorded this fungus form Eastern Himalayas as *Fomes geotropus* while Thind & Rattan (1972a) reported it from North Western Himalayas as *Fomes ulmarius*. Cartwright & Findlay (1958) do not consider the two as identical while Lloyd (1915a) suggest that *F. geotropus* may be a tropical form of *F. ulmarius*.

RIGIDOPORUS VINCTUS (Berk.) Ryv.

Norw. J. Bot. 19 : 139-144, 1972. Fig. 27. c-f., Pl. , fig.

Polyporus vinctus Berk. Ann. Mag. Nat. Hist. 9 : 196, 1852.

P. carneopallens Berk., Hook. J. Bot. 8 : 235, 1856.

P. benetostus Berk., J. Linn. Soc. 16 : 52, 1878.

Poria hyposclera Berk., Grevillea 10 : 102, 1882.

Polyporus rufolateritius Kalch., Grevillea 10 : 104, 1882.

Poria vincta (Berk.) Cke., Ibid. 14 : 110, 1886.

Polyporus porphyrophaea Bres., Hedwigia 56 : 296, 1915.

P. hypobrunnea Petch, Ann. Roy. Bot. Gard. Peradeniya 6 : 137, 1916.

Fruitbody annual to perennial, resupinate, adnexed and inseparable from host surface, smooth and even while fresh, rough and with some irregular splits when dry, up to 2mm thick at the centre, receding to wards margin. Pore surface flesh (6B3) to orange white (5A2), whitish towards margin, lighter when dry, adpressed and

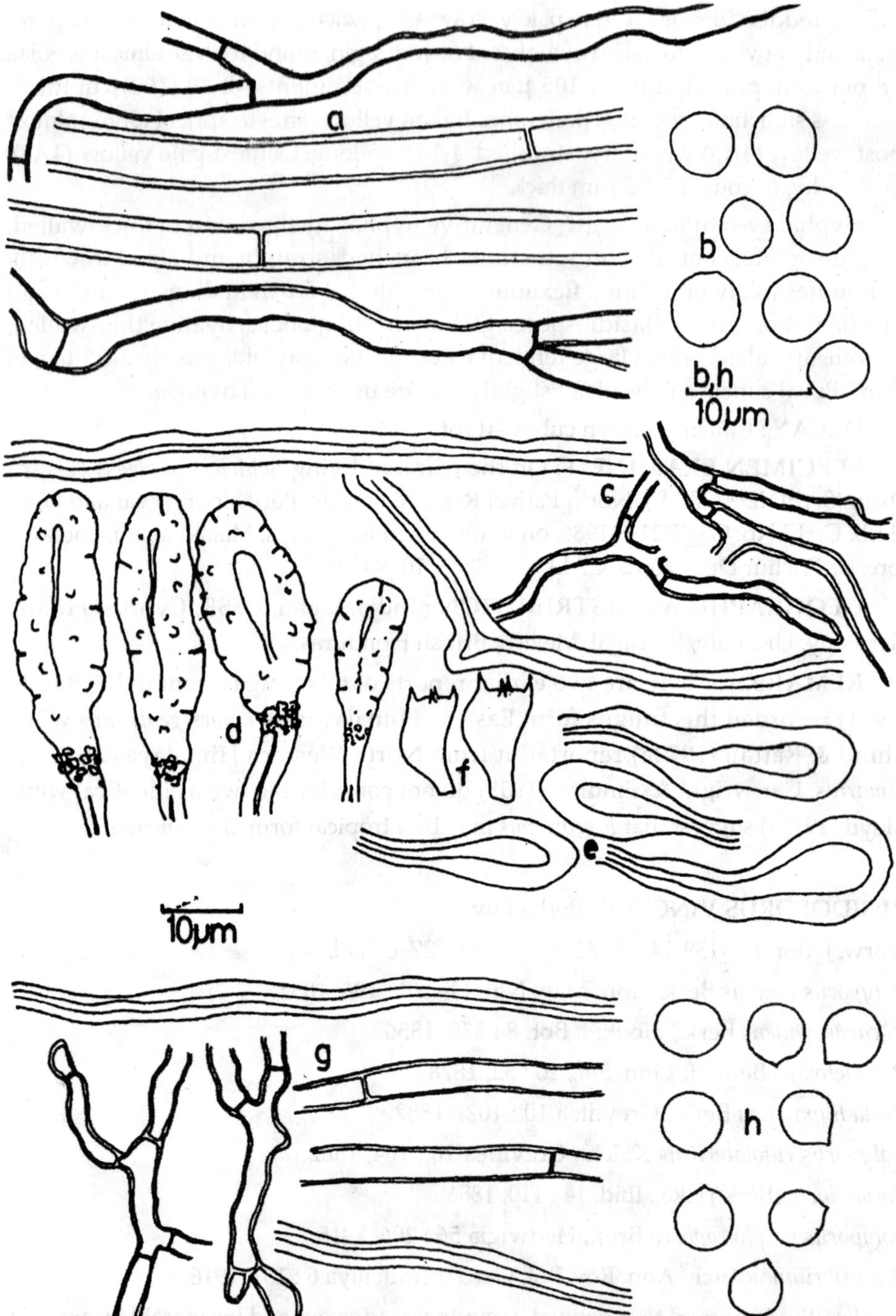

Fig. 27 : a-b *Rigidoporus microporus* : a - generative hyphae ; b - basidiospores . c - f *Rigidoporus vinctus* : c - generative hyphae; d - cystidia; e - cysidioles; f - basidia. g - h *Rigidoporus ulmarius* : g - generative hyphae; h- basidiospores.

spreading; pores minute and not visible to naked eye, absent towards margin, regular, 8-10 per mm; pore mouth 60-90 μm broad ; dissepiments 40-60(90) μm thick; pore tubes somewhat concolorous with the pore surface, sunk into equal depths, upto 1.5 mm long, old tubes invaded by context-tissue. Context yellowish grey (4B2), 180-250 μm thick at the centre, thinning considerably towards margin.

Hyphal system monomitic. Generative hyphae hyaline to pale-coloured, thick-walled with narrow but distinct lumen, simple-septate, 4-5 mm broad. Cystidia numerous, densely encrusted, thick-walled, cylindric to clavate, 20-70 × 11-16.5 μm; cystidioles hyaline, thin-walled, clavate to obpyriform, 15-20 × 7.5-10 μm. Basidia broadly clavate, 2 or 4-spored, 10-14.5 6-7 × μm; sterigmata short and stout. Basidiopores not observed.

DECAY : White fibrous rot.

SPECIMEN EXAMINED : On decaying logs of *Tectona grandis* Linn. f. (VERBENACEAE), Kanjirakadavu, Nilambur Forest Division, Malappuram Dist., Herb. CALI No. G156, 7-3-1985.

GEOGRAPHICAL DISTRIBUTION : North and South America. Europe, New Zealand, India, Sri Lanka, Philippines, Cook Island, Africa, French Antilles and Guiana.

REMARKS : This polypore was earlier known from India as *Poria vincta* (Bakshi, 1971). The present collection agrees with the description of this polypore from North America (Lowe, 1985), New Zealand (Cunningham, 1965) and East Africa (Ryvarden & Jonansen, 1980). Rajchenberg (1984) describes *R. vinctus* from Argentina as having a dimitic hyphal system *R. vinctus* was collected only once form this region, causing decay in teak.

SKELETOCUTIS Kotl. & Pouz.

Ceska Mykol. 12 : 103, 1958.

Fruitbody annual, pileate or resupinate. Pileus surface white, finely tomentose, thin. Pore surface first white then orange, gelatinous and dense while fresh; tubes concolourous with pore surface, distinctly delimited towards a white and thin context or subiculum. Hyphal system dimitic, rarely trimitic; generative hyphae thin-walled and with clamps; skeletal hyphae hyaline and thick-walled. Cystidia none. Basidiospores cylindrical to allantoid, smooth, hyaline, thin-walled and nonamyloid. Causing white rot.

Type species : *Skeletocutis amorpha* (Fr.) Kotl. & Pouz.

Only one species has been recorded from Kerala.

SKELETOCUTIS NIVEA (Jung.) Keller

Persoonia 10 : 347-355, 1979. fig. 28. a-f; Pl. , fig.

Polyporus niveus Jungh., Verh. Batav. Gennootsch. 17 : 48, 1838

P. chioneus Fr., Sensu Bres., Ann Mycol. 6 : 37, 1908, non Fr., 1821.

P. semisupinus Berk. & Cke., Grevillea 1 : 50, 1872.

P. semipileatus Peck, N. Y. State Mus. Ann. Rep. 34 : 43, 1881.

Tyromyces chiones (Fr.) Karst., Rev. Mucol. 3 : 17, 1881.

Leptoprus chioneus (Fr.) Quel., Fl. Mycol. P. 385, 1888.

Tyromyces semipileatus (Peck) Murr., North Amer. Fl. 9 : 35, 1907.

Leptoporus semipileatus Pilat, Atl Polyp. P. 183, 1938.

Poria coprosmae G. H. Cunn., N. Z. Dep. Sci. Industr. Res. Bull. 72 : 18, 1947.

Leptotrimitus semipileatus (Peck.) Pouzar, Ceska Mykol. 20 : 175, 1966.

Incrustoporia nivea (Jungh.) Ryv., Norw. J. Bot. 19 : 322, 1972.

Fruitbody annual, resupinate to effused-reflexed, narrow and elongated, imbricate, flexible while fresh, rigid when dry, 3-10 × 2-3 × 0.1-0.4 cm. Pileus surface white or with a pale yellow (3A3) tinge, pale yellow when dry, finely matted tomentose, azonate; margin sometimes inflexed and smooth. Pore surface white, Pale yellow (3A3) or slightly darker when dry, uneven; pores small but visible to naked eye, decurrent, absent near margin, irregular; pores round to oval, 4-6 per mm; pore mouth 110-180(200) μm wide; dissepiments 35-85 μm thick; pore tubes white, straight or oblique, sunk into equal depths, shallow, up to 1mm long . Context white, less than 2mm thick, homogeneous.

Hyphal system trimitic. Generative hyphae hayline, slightly thick-walled, branched, septate with clamps, uncommon, 2-2.5 μm in diam. Binding hyphae hyaline, thick-walled with obiterate lumen or solid, well branched, sinuous and flexuous, non-septate, 1.75-3.5 μm in diam. Skeletal hyphae hyaline, thick-walled with lumen broad and distinct, unbranched or rarely branched (terminally), non-septate, mostly cylindrical, upto 6 μm in diam. Encrusted hyphal tips present in dissepiments. Basidia clavate, 4-spored, 10-12 × 3-4.5 μm Basidiospores very small, hyaline, allantoid, smooth, egattulate, nonamyloid, 2-3 × 0.5 μm.

DECAY : Associated with white fibrous rot, usually on decorticated branchlets.

SPECIMEN EXAMINED : On decaying logs of *Anacardium occidentale* Linn. (ANACARDIACEAE), Botanical garden, Calicut University Campus, Malappuram Dist., Herb. CALI No. G71, 6-7-1984, Gl05, 1-6-1985; on fallen branches of *Grewia tiliifolia* Vahl (TILIACEAE), Vengoli peak, Teak Plantation Division, Parambiculam, Palghat Dist., Herb. CALI No. G207, 11-7-1985; on unidentified forest logs, Kadavakkunnu, Nadugani Forests, Nilambur, Malappuram Dist., CALI No. Gl41, 13-11-1984.

GEOGRAPHICAL DISTRIBUTION : Cosmopolitan.

REMARKS : *S. nivea* has earlier been recorded from India by Bakshi (1971) as *Polyporus semipileatus.* The present collection agrees closely with the description of *S. nivea* by Bourdot & Galzin (1927), Klan & Kotilova (1982), Lowe (1966) and Lowe & Gillbertson (1961). Bonderstev (1953), Ireneia (1980), Niemela & Uotila (1977), Overholts (1953) Ryvarden (1978), and Ryvarden & Johansen (1980) reports smaller pores. David (1982) monographed this genus and remarked on the varied hyphal system of *S. nivea.* This polypore can be easily identified by its white, effused-reflexed fruitbodies with a narrow pileus, small pores and minute allantoid spores.

TRAMETES Fr.

Fl. Scan. P. 339, 1835.

Fruitbody annual or at least short-lived, pileate, broadly attached to distinctly dimidiate, solitary or imbricate, tough when fresh, remaining mostly so when dry. Pileus surface glabrous to hirsute, white, cream, grey or unevenly pale brown. Pore surface light-coloured, white, cream to grey, become pale brown to ochraceous with age; pores round to angular, more rarely radially elongated, entire, in some species dentated when old, but never distinctly deadaleoid or lamellate, hyphal pegs absent or present, tubes non-stratified. Context tough, white to cream or pale-straw-coloured, in some species with a thin black zone between the context and the upper tomentum. Hyphal system trimitic; generative hyphae thin-walled and with clamps, binding and skeletal hyphae hyaline, thick-walled to solid, usually abundantly present. Cystidia none. Basidiospores, ellipsoid, cylindrical to allantoid, smooth,thin-walled and nonamyloid. Mostly on deciduous wood, Cosmopolitan genus.

Type genus : *Trametes suaveolens* (Fr.) Fr.

Eight species have been recorded from Kerala.

KEY TO SPECIES

1. Pileal surface velutinate, hirsute, or strigose 2
1.′ Pileus surface glabrous 5
2. Pileus surface finely tomentose, glabrescent when mature, White to cream-coloured .... *T. cotonea*
2.′ Pileus surface adpressed-velutinate to strgose or agglutinated, orangish white to brownish 3
3. Pileus surface pale-coloured, azonate or faintly Zonate, adpressed-velutinate .... *T. lactinea*
3′. Pileus surface pale grey to brownish, azonate to deely zonate; tomentose to strigose or hairs agglutinated 4
4. Pileus surface pale grey to brown, deeply striate, tomentose to strigose; pores 3-4 per mm .... *T. hirsuta*
4′. Pileus surface brown, hairs agglutinated to appear warty; pores 2-3 per mm .... *T. modesta*
5. Hyphal pegs present; hymenial surface daedeleoid to poroid; pores 2-3 per mm .... *T. scabrosa*
5.′ Hyphal pegs absent; hymenial surface poroid; pores 4-8 per mm 6
6. Pileal surface partly dark brown to soot brown; stipe rundimentary or attached with a converging narrow base .... *T. cingulata*
6.′ Pileus surface white to yellowish; stipe absent, attached with a broad base 7
7. Pileus surface yellowish; pores 6-8 per mm; spores $3\text{-}3.5 \times 2\text{-}2.5$ μm .... *T. marianna*
7′. Pileus surface whitish; pores 4-5 per mm; spores $5\text{-}6.5 \times 2\text{-}3$ μm .... *T. varians*

TRAMIES CINGULATA Berk.

Hook. J. Bot. 6 : 164, 1854. fig. 27. g-j., Pl. , fig.

Fruitbody annual, pileate, laterally attached with a reduced stipe, solitary, semi-circular to flabelliform, applanate, coriaceous while fresh, hard when dry; 1.9-6.5 × 1.7-4.5 × 0.2-0.7 cm. Pileus surface yellowish white (3A2) with a greenish tinge to bronze (5E5) or soot brown (5F5), rarely dark grey (1F1), yellowish white (3A2) near margin, colour often in patches; smooth but slightly uneven, glabrous, concentrically striate; margin thick, smooth, often irregularly wavy. Stipe reduced, concolourous with the pileus surface, solid, inside tissue white. Pore surface yellowish white (2A2) to somewhat cream (4A3), pores small, visible to naked eye, absent near margin, never decurrent, pores round to ovate, 5-6 per mm; pore mouth 100-135 μm wide, shallow towards margin; disspiments 65-100 μm wide; pore tubes concolorous with the pore surface, sunk into almost equal depths, straight, 1-2 mm long. Context concolorous with pore tubes, light yellow (4A4) when dry, homogeneous, 1-5 mm thick.

Hyphal system trimitic. Generative hyphae hyaline, thin-walled to fairly thick-walled, usually simple-septate, rarely clamped, 2-3 μm in diam. Binding hyphae hyaline, thick-walled with obliterate lumen or solid, branched, less frequent, 2-3.5 μm in diam. Skeletal hyphae hyaline to pale-coloured, thick-walled with broad and distinct lumen, flexuous or straight, 3-5.6 μm in diam. Cystidia absent. Basidia not observed. Basidiospores hyaline, thin-walled, ovate apiculate, smooth, eguttulate, nonamyloid, 5-6.9 × 3.5-4 μm; spore print white.

DECAY : White rot.

SPECIMEN EXAMINED : On decaying branchlets of *Terminalia paniculata* Roth (COMBRETACEAE), below Vengoli Peak, Tunacadavu, Parambiculam forest, Teak Plantation Division, Palghat Dist., Herb. CALI No. G266, 12-11-1985; unknown host, Vattilapara, near Vazhachal, Sholayar forests, Trichur Dist., Herb. CALI No. G 84 26-7-1984, Pathenthode, Silent Valley National Park, Palghat, Herb CALI No. G34 7-9-1983, Botanical Garden, Calicut University Campus, Malappuram Dist. Herb. CALI No. G362, 23-1-1987.

GEOGRAPHICAL DISTRIBUTION : Afro-Asian species.

REMARKS : *T. cingulata* has been reported as early as 1854 from India by Berkeley and later by Bakshi (1971). Ryvarden & Johansen (1980) has reported it from East Africa. This is the first report of *T. cingulata* from south India.

TRAMTES COTONEA (Pat.& Har.) Ryv.

Norw. J. Bot. 19 : 236, 1972. fig. 28. k-m., Pl. , fig.

Polystictus cotoneus Pat. & Har., Bull. Soc. Mycol. Fr. 9 : 208, 1893.

Polyporus cotoneus (Pat. & Har.) Sacc., Syll. Fung., 11 : 93, 1895.

Fruitbody annual, sessile, attached with a broad lateral base, effused- reflexed to pileate, solitary, sometimes imbricate, dimidiate to flabelliform, applanate, corky while fresh, tough when dry; 2.5-5.5 × 2-3 × 0.3-0.6 cm. Pileus surface white to cream (4A3) while fresh, pale yellow (3A2) when dry, finely tomentose, glabrescent when old, faintly radially wrinkled when dry; margin entire, thin, smooth rigid. Pore surface pale yellow (4A3), slightly darker when dry; pores absent near margin, large and conspicuous to naked eye, angular, 2-3 per mm, almost regular; pore mouth 250-500 μm wide; dissepiments 100-150 μm thick; pore tubes concolorous with the pore surface, 1-3 mm long, sunk into unequal depths. Context white, cottony, tapering slowly, towards margin, 0.5-3 mm thick.

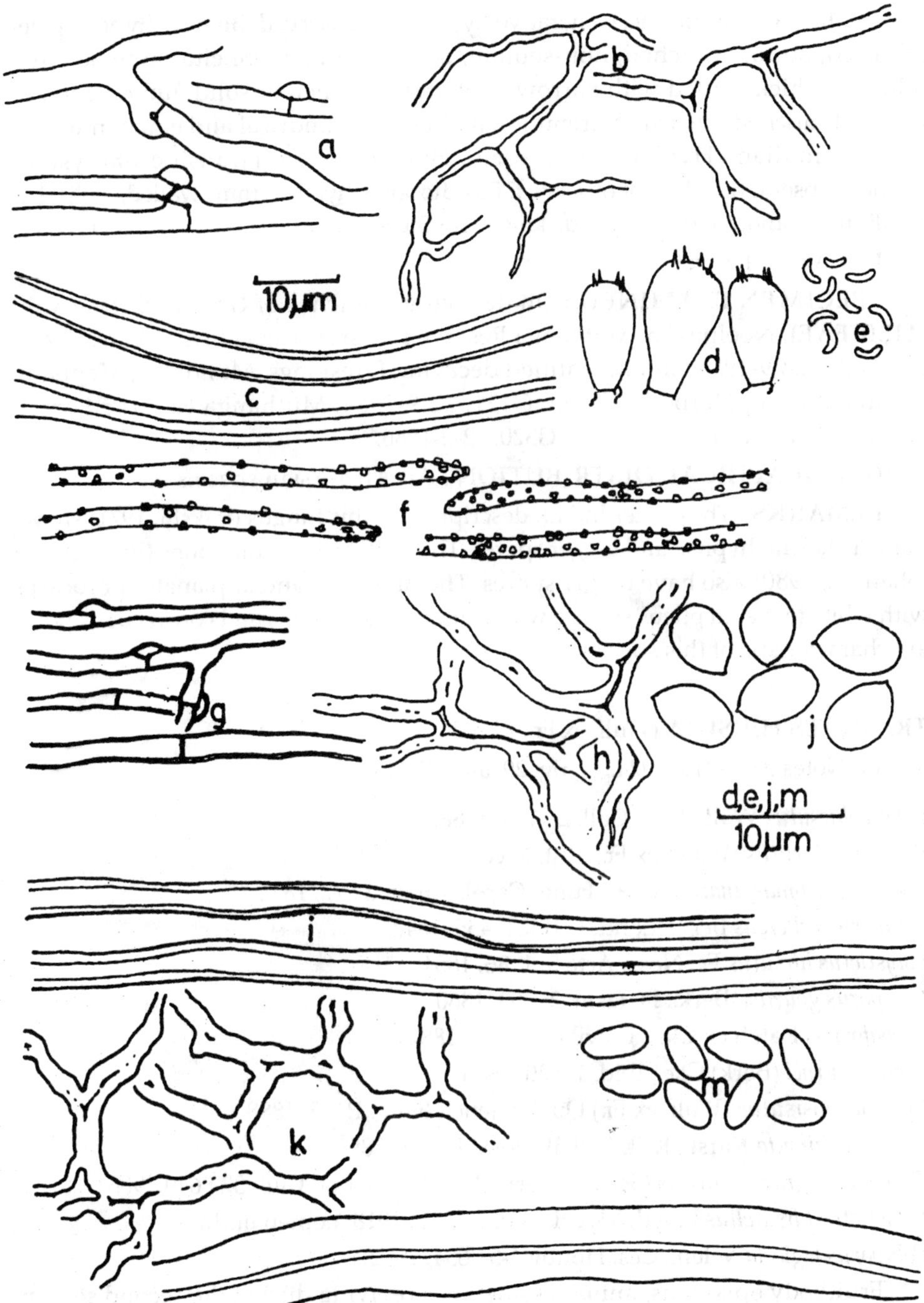

Fig. 28 : a-f *skeletocutis nive* a : a - generative hyphae; b - binding hyphae; c - skeletal hyphae; d - basidia; e - basidiospores; f - encrusted hyphae at pore mouth.

g-j *Trametes cingulata* : g - generative hyphae; h - binding hyphae; i - skeletal hyphae; j - basidiospores.

k - m *Trametes cotonea* : k - binding hyphae; l - skeletal hyphae; m - basidiospores.

Hyphal system trimitic. Generative hyphae not observed. Binding hyphae pale-coloured, solid, branched, non-septate, 2-3 µm in diam. Skeltal hyphae pale-coloured, thick-walled with narrow to obliterate lumen or solid, unbranched or weakly branched towards extremities, main stem cylindrical and equal in width, 5-10 µm in diam. Hyphal pegs present. Hymenium with translucent crystals. Cystidia absent. Basidia not observed. Basidiopores hyaline, thin- walled, cylindric to elliptical, smooth, nonamyloid, 4.5-6 × 2-5-3 µm.

DECAY : White rot.

SPECIMEN EXAMINED : On decaying branches of *Grewia tiliifolia* Vahl **(TILICEAE)**, Noolpuzha, Muthanga Reserve Forest, Wyanaad Dist., Herb. CALI No. G319, 22-9-1986; on unidentified decaying forest logs, Mundakai, Meppadi, Wyanaad Dist., Herb. CALI No. G121, 22-9-1984; Muthanga Reserve Forest, Wyanaad Dist., Herb. CALI No. G320, 22-6-1986.

GEOGRAPHICAL DISTRIBUTION : An Afro-Asian species.

REMARKS : The earlier Indian description of this fungus (Bakshi, 1971) shows thinner skeletal hyphae and bigger spores. The East African collections (Ryvarden & Johansen, 1980) also have bigger spores. The dimidiate and applanate sporocarps with white to cream pileus surface with concentric zonations and conspicuous pores are characteristic of this species.

TRAMETES HIRSUTA (Wulf. ex Fr.) Lloyd

Mycol. Notes 73 : 1319, 1924. fig. 29. a-e., Pl. , fig.

Boletus hirsutus Wulf., Jacq. Coll. 2 : 149, 1788.
Polyporus hirutus (Wulf.) ex Fr., Syst. Mycol. 1 : 367, 1821.
Boletus nigromarginatus Schw., Fung. Carol. Super. 2 : 72, 1822.
Polyporus vellereus Berk., Lond. J. Bot.1: 455, 1842.
Polystictus hirsutus Fr., Nova Acta 3, 1: 86, 1851
Ployporus gourliei Berk., FI Tasm. 2: 253, 1860.
P. pisiformis Katch.ex Cke., grevillea 10 : 98,1882.
Fomes gourliei (Berk) Cke., Ibid. 14: 20, 1885.
Coriolus hisrsutus (Wulf. ex Fr.) Quel., Elench Fung. P.175, 1888
Hansenia hirsuta Karst., Krit. Finl. Basidsv. P.304, 1889.
Coriolous nigromarginatus (Schw.) Murr., Bull. Torrey Bot. Club 32 : 649,1906
Polystictus hirsutellus Lloyd, Mycol. Letter 67 : 2, 1918, nomen nudum.
Polyporus fagicola Velen., Ces. Houby 4-5 : 654, 1922.

Fruitbody epixylous, annual, sometimes reviving during the second season, pileate, sessile, sometimes effused-reflexed, imbricate and confluent, rarely solitary, attached laterally with a broad base, semi-circular to flabelliform, applanate, coriaceous and flexible while fresh, tough when dry; 3-12 × 2.2-11 × 0.21 cm. Pileus surface clay (5DS) to mustard brown (5E6) or pale grey (IBI), greenish grey (28B2) to greyish green (28B3) towards margin, tomentose, hispid or strigose, concentrically zonate while young, deeply striate in large ones, margin thin and entire. Pore surface pale yellow (3A2), wine yellow (3B3), yellowish white (3A2) or yellowish grey (3C2), even and smooth; pores large and visible to naked eye, extending almost to the margin, oval to round, regular, 3-4 per mm; pore mouth 150-210 μm broad; dissepiments 90-150(210) μm thick; pore tubes concolourous with the pore surface, sometimes white to pale yellow (3A2), 2-3 mm long, straight, thinning towards margin, sunk into unequal depths. Context yellowish white (3A2) to concolorous with the pileus surface, homogeneous while young, sometimes duplexed when old, upper layer (towards crust) thick and dark brown or concolorous with pileus surface and lower layer (towards pores) yellowish white or greyish white, 0·2-0·8 cm thick.

Hyphal system trimitic. Generative hyphae hyaline, fairly thick-walled, branched, septate with clamps, 1.75-2.5 μm in diam. Binding hyphae hyaline, thick-walled with narrow lumen or solid, moderately to well branched, non-septate, 3-5 μm in diam. Skeletal hyphae hyaline or pale-coloured, thick-walled with narrow lumen, non-septate, unbranched, extremities twisted and with short side branches, 4.5-6 μm in diam. Basidia long, clavate, 15-17(21) × 3.5-4 μm; sterigmata up to 4 μm long. Cystidia absent. Hyphal pegs present. Basidiospores hyaline, oblong-elliptical to slightly reniform, sometimes ovate, eguttulate, nonamyloid, 4.5-6.5 × 2-3 μm.

DECAY : White fibrous rot.

SPECIMEN EXAMINED : On decaying logs of *Artocarpus heterophyllus* Lam. (MORACEAE), Calicut University Campus, Malappuram Dist., Herb. CALI No. G60, 19-6-1984; on dead stem of *Acacia auriculiformis* A. Cunn. ex Benth. (MIMOSACEAE), Calicut University Campus, Malappuram Dist., Herb. CALI No. G 278, 20-7-1985; on dead stems of *Phyllanthus acidus* (L.) Skeels (EUPHORBIACEAE), Thiruvambady, Trichur Dist., Herb. CALI No. G329, 24-8-1986; on branches of *Zanthaxylum rhetsa* (Roxb.) D.C. (RUTACEAE), Kanjirakadavu, Nilambur Forest Division, Malappuram Dist., Herb. CALI No. G158, 7-3-1985; unknown host, on decaying stumps, Karadimala, Brahmagiri Ranges, Tirunelli, Wyanaad Dist., Herb. CALI No. G259, 2-10-1985; Mundakai, Meppadi, Wyanaad Dist., Herb. CALI No. G243, 12-9-1985; Poothole, Trichur Dist., Herb. CALI No. G65, 1-7-1984; Kuppadi, Sultan Battery, Wyanaad Dist., Herb. CALI No. G293, 20-6-1986; Silent Valley National Park, Palghat Dist., Herb. CALI No. G43, 15-1-1984.

GEOGRAPHICAL DISTRIBUTION : World wide, not reported from Africa.

REMARKS : The present collection agrees remarkably well with the earlier Indian collections of *T. hirsuta* (Bose, 1937a; Bakshi, 1971; De & Roy, 1980). The description of *T. hirsuta* by Bourdot & Galzin (1927) shows bigger spores. The present collection appears similar to the collections from North America (Neuman, 1914; Lowe, 1942; Overholts, 1953), Britain (Rea, 1922), Russia (Boundarstev, 1953), Czechoslovakia (Kotlaba, 1977), North Europe (Ryvarden, 1978) and Argentina (Rajchenerg, 1982). The absence of this common tropical polypore in Africa is noteworthy.

TRAMETES LACTINEA (Berk.) Pat.

Essai Tax. P. 92. 1900 Fig. 29, f-h., Pl. , fig.

Polyporus lactineus Berk., Ann. Nat. Hist. 10 : 373, 1843.

Trametes levis Berk., Lond. J. Bot. 6 : 507, 1847.

Coriolus lactineus (Berk.) G.H. cunn., Proc. Linn. Soc. N.S.W. 75 : 229, 1950.

Fruitbody annual, pileate, sessile, attached with a broad lateral base, sometimes effused-reflexed, easily separable from the host, highly imbricate and confluent, dimidiate to semi-circular, triquetrous to broadly applanate, corky while fresh, tough when dry; 4.5-10 × 2.5-6 × 0.3-1.1 cm. Pileus surface yellowish white (3A2) to orange white (5A2), sometimes slightly brown towards base, yellowish white (3A2) towards margin, uneven and rough, adpressed-velutinate, faintly zonate concentrically, radially grooved, slightly crusty-nodulose towards base; margin thick, smooth and rounded. Pore surface yellowish white (3A2) with a greyish tinge when collected, cream (4A3) coloured on brusing, pale yellow (3A3) to brownish orange (5B4) when dry; pores extending almost up to the margin, large and visible to naked eye, oval to round, almost regular, 2-3 per mm; pore mouth (200) 250-310 μm wide; dissepiments 95-185 μm thick; pore tubes concolorous with the pore surface or marble white (5B2) towards the context, 2-5.5 mm long, sunk into unequal depths. Context cream (4A3) coloured, fibrous, homogeneous, 1-5 mm thick.

Hyphal system trimitic. Generative hyphae hyaline, thin-walled, branched, septate with clamps, well represented in dissepiments, rare in context, 2.5-4.5 μm in diam. Binding hyphae hyaline to slightly pale-coloured, profusely branched, sometimes with numerous short side branches, usually solid, common in context, 3-4 μm in diam. Skeletal hyphae hyaline to yellowish, unbranched, non-septate, thick-walled to solid with obliterate or definite lumen, 4-5 μm in diam. Cystidia absent. Basidia and basidiospores not observed.

DECAY : White stringy rot.

SPECIMEN EXAMINED : On decaying logs of *Dalbergia latifolia* Roxb. (FABACEAE), Karadimala, Brahmagiri Ranges, Tirunelli, Wyanaad Dist., Herb. CALI No. G 261, 2-10-1985.

GEOGRAPHICAL DISTRIBUTION : Sri Lanka, India, Pakistan, Philippines, New Zealand, Australia, Fiji, New Guinea, West Indies, South Africa.

REMARKS : This polypore has earlier been reported from India (Bose, 1934; Anonymous, 1950). The present collection appears similar in macro-and microcharacters with Bakshi's (1971) report except for the extremely large size of the fruitbody. The reddish colouration of the pore surface developed on bruising reported by Lloyd (1904-1991) is not observed here. The African fruitbody (Ryvarden & Johansen, 1980) have smaller pores and slightly broader hyphae. The thick, yellowish white fruitbodies with soft velutinate pileus surface and conspicuous pore are valuable field characters for this polyore.

TRAMETES MARIANNA (Pers.) Ryv.
Persoonia 7 : 309, 1973. fig. 29. i-m., Pl. , fig.

Polyporus mariannus Pers. In Gaudichaud, Voy. Aut. Monde, Paris p. 173, 1827.
P. paleaceus Fr., epicrisis p. 471, 1838.
Trametes paleacea Fr., Nova Acta 3, 1 : 97, 1851.
T. sprucei Berk., Hook. Lond. J. Bot. 8 : 236, 1856.
Polyporus proteus Kalch. ex Cke., Grevillea 10 : 102, 1882, non Berk, 1843.
P. palliseri Berk. ex Cke., Ibid. 10 : 89, 1882.
Polystictus proteiformis Cke., Ibid. 14 : 81, 1886.
Polyporus argentatus Cke., Ibid. 15 : 20, 1886.
Coriolus paleaceus (Fr.) G.H. Cunn., Proc. Linn. Soc. N.S.W. 75 : 234, 1950.
Trametes proteiforma (Cke.) G.H. Cunn., Ibid. 75 : 238, 1950.

Fruitbody epixylous, annual, effused-reflexed to pileate, sessile, attached laterally with a broad base, solitary or imbricate, confluent, semi-circular to dimidiate, coriaceous while fresh, hard when dry; 5-8 × 2.5-4 × 0.5 cm. Pileus surface butter yellow (4A5) to light orange (5A5), pale yellow (3A3) towards margin, glabrous, striations, concentric and more prominent towards periphery, faintly crusty towards the base; margin rigid and entire. Pore surface pale yellow (3A3) while fresh, grey orange (5B4) when dry; pores minute and not visible to naked eye, not extending to the margin, almost angular to oval to round, 6-8 per mm; pore mouth 75-100 μm wide; dissepiments 25-60 μm thick; pore tubes concolorous with the pore surface, up to 2.5 mm long, straight, sunk into almost unequal depths. Context pale yellow (4A3), homogeneous, up to 2.5 mm thick.

Hyphal system trimitic. Generative hyphae hyaline, thin-walled, septate with clamps, sparsely branched, confined mostly to dissepiments and younger parts, 2.25-3 μm in diam. Binding hyphae yellowish, thick-walled with narrow to broad lumen, highly branched, non-septate, 1.5-4.5 μm in diam. Skeletal hyphae yellowish, thick-walled with distinct lumen, unbranched, non-septate or rarely pseudoseptate, cylindrical and straight, 4.5-6 μm in diam. Basidia clavate, almost 11 × 5 μm, 4-spored; sterigmata small and straight, up to 1 μm long. Cystidia absent. Basidiospores hyaline, thin-walled, ellipsoid to slightly ovoid, eguttulate, smooth, 3-3.5 × 2-2.5 μm.

DECAY : White rot.

SPECIMEN EXAMINED : On dead branches, host not identified, Mundakai forests, Meppadi, Wyanaad Dist., Herb. CALI No. G242, 12-9-1985.

GEOGRAPHICAL DISTRIBUTION : Africa, Samoa, New Zealand, India.

REMARKS : Bakshi (1971) following the description of Bose (1922a) included this polypore among the inadequately known *Trametes* from India. He reported bigger spores than those of the present collection. Cunningham (1956) describes *T. paleacea* and *T. mulleri* which are synonyms of *T. marianna* (Ryvarden, 1973) as separate species, and both of them show wider pore mouth and variable spores. The east African fruitbodies (Ryvarden & Johansen, 1980) seem closer to the present collection. The sporophores of *T. marianna* superficially resemble young fruitbodies of *T. cingulata,* however the latter become dark greyish with age.

TRAMETES MODESTA (Fr.) Ryv.

Norw. J. Bot. 19 : 236, 1972. fig. 30. a-c., Pl. , fig.

Polyporus modestus Fr., Linnaea 5 : 519, 1830.

P. atypus Lev., Ann. Sci. Nat. 3, 2 : 184, 1844.

Fruitbody annual to biennial, effused-reflexed to pileate, sessile, imbricate and confluent, attached with a broad lateral base; semi-circular to dimidiate, often conchate, applanate, thick and coriaceous while fresh, rigid when dry; 2-4.5 × 1.5-3 × -1 cm. Pileus surface brownish orange (7C3-7C5), Somalis (7E5) towards base, sunburn (6D5) when dry; almost even, slightly rough, azonate, context hyphae protuding on the pileus surface agglutinated to appear rough and warty; margin smooth, rounded, entire, rigid when dry. Pore surface white with a greenish shade and with a pinkish margin while fresh, light yellow (4A4) when dry, uneven, sometimes with small pits or depressions; pores not extending to the margin, small and not visible to naked eye, round, regular, 5-6 per mm; pore mouth 90-150 µm wide; dissepiments 25-40 µm thick; pore tubes concolorous with the pore surface, stratified, 2-4 mm long in each stratum, sunk into somewhat unequal depths. Context white with an orange tinge, orange white (5A2) when dry, homogeneous.

Hyphal system trimitic. Generative hyphae hyaline, thin-walled, occasionally branched, septate with clamps, easily breaking, moderately represented in all tissues, 2.5-5 µm in diam. Binding hyphae hyaline, thick-walled with obliterate lumen or almost solid, branched, non-septate, 2.5-4 µm in diam. Skeletal hyphae hyaline to pale-coloured, thick-walled, with narrow to obliterate lumen, usually unbranched, non-septate, central stem cylindrical and straight, 3-7.5 µm in diam. Basidia and basidiospores not observed.

SPECIMEN EXAMINED : On decaying wood, host not identified, Kaduvakkunnu, Nadugani forests, Nilambur, Malappuram Dist., Herb. CALI No. G133, 13-11-1984; Silent Valley National Park, Palghat Dist., Herb. CALI No. G226, 15-9-1985.

GEOGRAPHICAL DISTRIBUTION : Pantropical, South America, Africa (Kenya and Burundi), South-East Asia (Thailand), India.

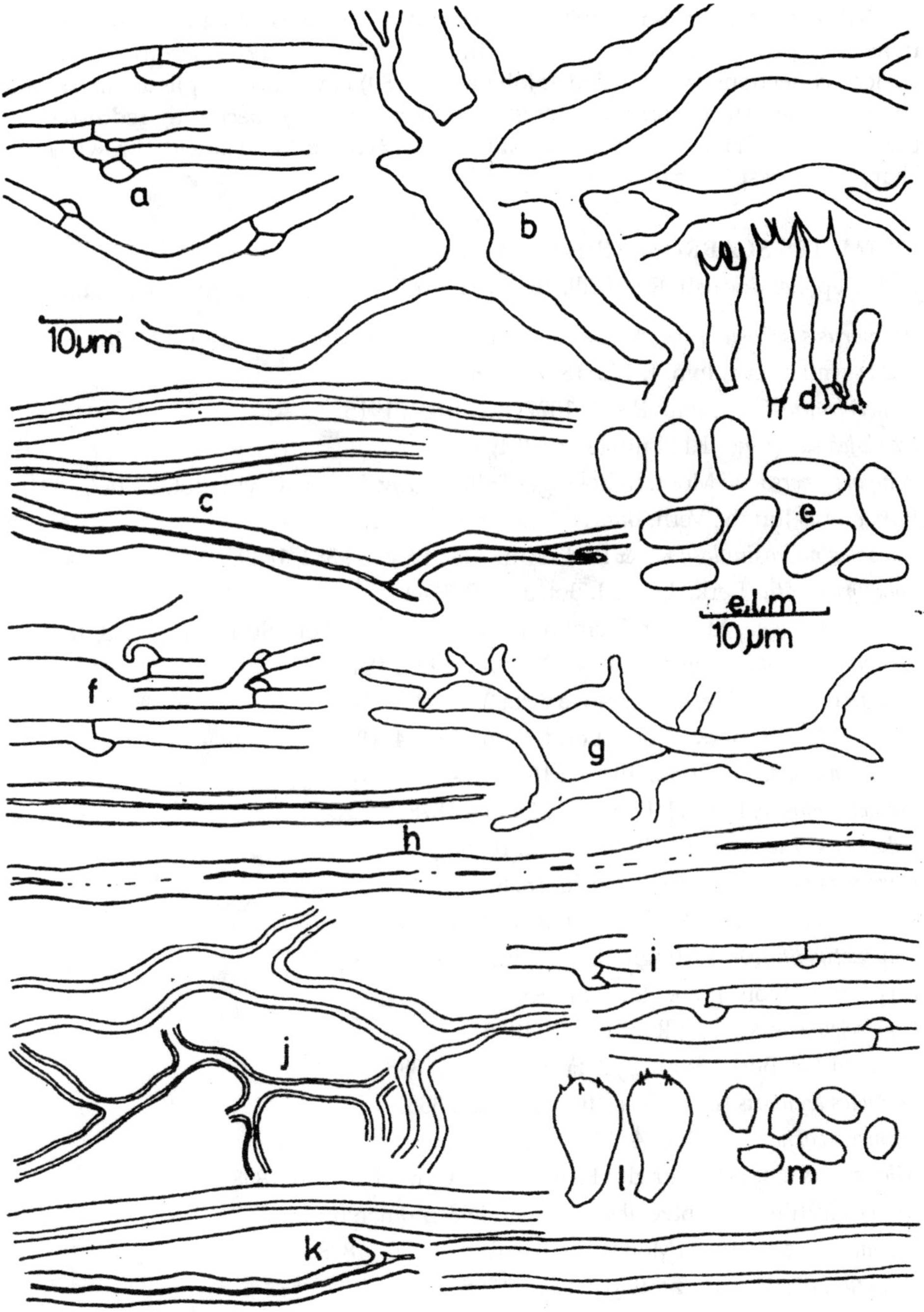

Fig. 29 : a-e *Trametes hirsuta* : a - generative hyphae; b - binding hyphae; c -skeletal hyphae; d - basidia; e - basidiospores.

f - h *Trametes lactinea* : f-generative hyphae; g - binding hyphae; h - skeletal hyphae.

i - m *Trametes marianna* : i - generative hyphae; j - binding hyphae; k - skeletal hyphae; l - basidia; m - basidiospores.

REMARKS : This is the first Indian record of *T. modesta* and it was collected only twice. The present collection agrees with Lloyd's (1904-1919) description while the African collections of Ryvarden & Johansen (1980) have variable pileus surface and smaller pores. The context when fresh has an orange tinge, becoming reddish when treated with KOH; in the dried context the reddish colour soon fades to brown. These characters are specific to this species of *Trametes.*

TRAMETES SCABROSA (Pers.) G.H. Gunn.
N.Z. Dep. Sci. Industr. Res. Bull. 164 : 162, 1965. fig. 30. d-h; Pl. , fig.

Polyporus scabrosus Pers., In Gaudichaud, Voy. Aut. Monde, Paris p. 172, 1827.
P. corrugatus Pers., Ibid. P. 172, 1827.
P. fuscobadius Pers., Ibid. P. 172, 1827, non Bres., 1915.
Daedalea sanguinea Kl., Linnaea 8 : 481, 1833.
Polyporus persoonii Mont. In Belanger & Bory, Voy. Ind. or. Crypt. Bot. 2 : 147, 1834.
P. indecorus Jungh., Verh. Batav, Genootsch. 17 : 51, 1838.
Daedalea microsinulosa Kl. & Berl., Ann. Nat. Hist. 3 : 382, 1839.
Polyporus tostus Berk., Lond. J. Bot. 3 : 189, 1844.
Hexagona cruenta Mont. In Hombron & Jaquinot, Voy. Pole. Sud. Bot. 1 : 208, 1845.
Polyporus tegularis Lev., Ann. Sci. Nat. 3, 5 : 131, 1846.
Polystictus tostus (Berk.) Fr., Nova Acta 3, 1 : 94, 1851.
Trametes hookeri Berk., Hook. Lond. J. Bot. 6 : 164, 1854.
Daedalea emodensis Berk., Ibid. 6 : 166, 1854.
Trametes moselei Berk., J. Linn. Soc. 16 : 40, 1878.
Polyporus aruensis Berk., Ibid. 16 : 43, 1878.
Hexagona picta Berk., Ibid. 16 : 50, 1878.
Fomes scabrosus (Pers.) Cke., Grevillea 14 : 69, 1886.
Polystictus indecorus (Jungh.) Cke., Ibid. 14 : 82. 1886.
P. persoonii (Mont.) Cke., Ibid. 14 : 85, 1886.
P. moseleii (Berk.) Cje., Ibid. 14 : 85, 1886.
Fomes tostus (Berk.) Sacc., Syll. Fung. 6 : 203, 1888.
Trametes tegularis (Lev.) Sacc., Ibid, 6 : 343, 1888.
Trametes cornea Pat., Ibid. 4 : 17, 1890.
Microporus indicorus (Jungh.) Kuntze. Rev. Gen. Pl. 3 : 496, 1898.
M. moselei (Berk.) Kuntze. Ibid. 3 : 497, 1898, momen nudum.
Scendidium cruentium (Mont.) Kuntze, Ibid. 3 : 516, 1898.
S. pictum (Berk.) Kuntze, Ibid. 3 : 516, 1898.
Scindalma scabrosum (Pers.) Kuntze, Ibid. 3 : 519, 1898.
S. tostum (Berk.) Kuntze, Ibid. 3 : 516, 1898.
Polystictus rufocinerescens Henn., Bot. Jahrb. 25 : 501, 1898.
Trametes persoonii (Mont.) Pat., Essai Tax. P. 93, 1900.

Polyporus albo-luteus Rostr., Bot. Tidsskr. 24 : 359, 1902.

Earliella cubensis Murr., Bull., Turrey Bot. Club 32 : 479, 1905.

Polystictus albo-luteus (Rostr.) Sacc., Syll. Fung. 17 : 128, 1905.

Earliella corrugata (Pers.) Murr., Bull. Torrey Bot. Club 34 : 468, 1907.

Favolus sub-rigidus Murr., Bull. Torry Bot. Club 35 : 398, 1908.

Polystictus cubensis (Murr.) Sacc., Syll. Fung. 21 : 324, 1912.

Hexagona sub-rigida (Murr.) Sacc. & Trott, ibid. 21 : 349, 1912.

Trametes corrugata (Pers.) Bres., Hedwigia 51 : 316, 1912.

T. cruenta (Mont.) Bres., Ann. Mycol. 14 : 229, 1916.

Polystictus formosae Lloyd, Mycol. Writ. 5 : 632, 1917.

P. formosae Yasuda, Bot. Mag. Tokyo 31 : 170, 1917.

Fomitopsis corrugata (Pers.) Imaz., Bull. Tokyo Sci. Mus. 6 : 91, 1943.

Coriolus corrugatus (Pers.) G.H. Cunn., Proc. Linn. Sco. N.S.W. 75 : 222, 1950.

Fruitbody annual, sessile, broadly effused-reflexed or pileate, attached with a broad lateral base, flabelliform to dimidiate or semi-circular, applanate, coriaceous to corky while fresh, rigid and tough when dry; 2-10 × 1.5-5.5 × 0.2-0.5 cm. Pileus surface dark brown (9F6) to violet brown (10E6) to Somalis (6E5) towards base, cream (4A3) to yellowish white (4A2) elsewhere and near margin; concentrically zonate, radially wrinkled, usually crusty and uneven, glabrous; margin lobed or entire, rounded or acute. Pore surface cream (4A3) while young, putty (4B2) or champagne (4B4) when dry; pores extending to the margin but shallow, large and visible to naked eye, daedaleoid, irregular, poroid and regular towards margin, 2-3 per mm; pore mouth 250-350 μm wide; dissepiments 95-100 μm thick; pore tubes concolorous with the pore surface, sunk into equal depths, up to 2 mm long, straight. Context yellowish white (3A2), homogeneous, 1-3 mm thick.

Hyphal system trimitic. Generative hyphae hyaline, thin-walled, branched, septate with clamps, 2-3 μm in diam., easily breaking into bits. Binding hyphae hyaline to pale-coloured, thick-walled with narrow lumen, higly branched, zig-zag, non-septate, 2.5-4 μm in diam. Skeletal hyphae hyaline to pale-coloured, thick-walled with narrow lumen, aseptate, cylindrical and flexuous, usually unbranched or with short branched near extremities, 3.5-6(7.5) μm in diam, Basidia long, clavate, 2 or 4-spored, 12-15 × 4.5-6 μm; sterigmata up to 3.5 μm long. Cystidia none. Basidiospores hyaline, cylindrical to oblong-elliptical, smooth, thin-walled, (6) 8-10 × 2.5 -4(5) μm. Hyphal pegs and large crystals frequents in hymenium.

DECAY : White stringly rot.

SPECIMEN EXAMINED : On dead stumps of *Artocarpus heterophyllus* Lam. (MORACEAE), Botanical Garden, Calicut University Campus, Malappuram Dist; Herb. CALI No. G72, 12-7-1984; on fallen branches of *Hevea brasiliensis* (H. B. & K.) Muel. Arg. (EUPHORBIACEAE), Kumarakam Bird Sanctury, Kottayam Dist., Herb. CALI No. G73, 21-3-1984; near CPCRI Cocoa Garden, Tamarasseri, Calicut Dist., Herb. CALI No. G283, 19-5-1986; on logs of *Tectona grandis* Linn. f. (VERBENEACEAE), Cannoly's Teak Plantation, Nilambur, Malappuram Dist., Herb. CALI No. Gl03, 13-2-1985; Kuppadi, near Sultan Battery, Wyanaad Dist., Herb. CALI No. Gl89, 19-6-1985; on decaying wood, host not identified, Calicut University Campus, Malappuram Dist., Nilambur Malappuram Dist., Herb. CALI No. G45, G50, 28-2-1984; Kanjirakadavu, Nilambur forest Division, Malappuram Dist., Herb. CALI

No. G91, 16-9-1983, Kuppadi, Sultan Battery, Wyanaad Dist., Herb CALI No. G296, 20-6-1986; Divisional Forest Officer, Nilambur, Malappuram Dist., Herb. CALI No. G150, Gl51, 7-3-1985; Wyanaad Forests, Wyanaad Dist., Herb. FRI No. 5139, K. Bagchee, 26-4-1950.

GEOGRAPHICAL DISTRIBUTION : Pantropical.

REMARKS: An earliar Indian collection of *T. scabrosa* from Mussoorie Hills (Thind & Chatrath, 1960) shows thinner hyphae, longer basidia and broader spores while Bakshi's (1971) description agrees closely with the present collection. The Newzealandean sporophores (Cunningham, 1965) shows some variability in pore size and spores. The south American (Fidalgo & Fidalgo, 1966) and East African (Ryvarden & Johansen, 1980) description of *T. scabrosa* also agree with the present collection.

This is a common polypore in Kerala and has been collected from several localities from open areas to evergreen forests. It attacks a wide range of hosts. *T. Scabrosa* can be easily identified in the field due to its cream-coloured pileus with dark brown colouration towards base, and daedaloid to irregular pores.

TRAMETES VARIANS Van der Byl.

South Af. J. Sci. 18 : 281, 1922. fig. 30. i-m., Pl. , fig.

Polyporus griseus Bres., Ann. Mycol. 10 : 494, 1912 nomen illegit., non Peck 1874.

P. radiato-rugosus Bres., Ibid. 18 : 36, 1920 nomen illegit., non Berk., 1839.

P. durbalensis Van der Byl., South Af. J. Sci. 18 : 261, 1922.

P. pectunculus Lloyd, Mycol. Writ. 7 : 1315, 1924 nomen illegit, non Lev., 1846.

Fruitbody annual, pileate, sessile, attached laterally with a broad base, imbricate and confluent, semi-circular to dimidiate, applanate, conchate when dry, coriaceous while fresh becoming tough and woody when dry; 2-9 × 1.5-4.5 × 0.3-1 cm. Pileus surface white to cream (4A3) with a greyish tinge, yellowish white (2A2) when dry; slightly scrupose, azonate, radially striate, more wrinkled near margin, often slightly crusty; margin thin, entire and incurved. Pore surface white with a greenish tinge while fresh, pale yellow (3A2) or darker when dry, uneven and rough; pores small and visible to naked eye, absent near margin, round or vaguely angular, regular, 4-5 per mm; pore mouth 135-160 μm wide; dissepiments 75-100 (125) μm thick; pore tubes concolorous with pore surface, 3-4 mm long, sunk into almost equal depths, receding towards margin. Context white to cream (4A3), 3-4 mm thick, fibrous.

Hyphal system trimitic. Generative hyphae hyaline, thin to thick-walled, sparsely branched, frequently septate with clamps, 2.5-3.5 μm in diam. Binding hyphae hyaline, thick-walled to solid, arborescent, twisted and zig-zag, moderately present, 2-5 μm in diam. Skeletal hyphae hyaline to pale-coloured, thick-walled with obliterate lumen or solid, central stem unbranched and almost cylindrical to slightly flexuous, branched towards extremities, sparsely septate with clamps, broken at nodes when clamped, 4-6(7.5) μm in diam. Cystidioles hyaline, thin-walled, ventricose to legeniform, 20-25 × 4.4-7.5 μm. Basidia not observed. Basidiospores hyaline, cylindrical to sub-fusiform, thin-walled, smooth, eguttulate, nonamyoid, 5-6.4 × 2-3 μm.

DECAY : White pocket rot.

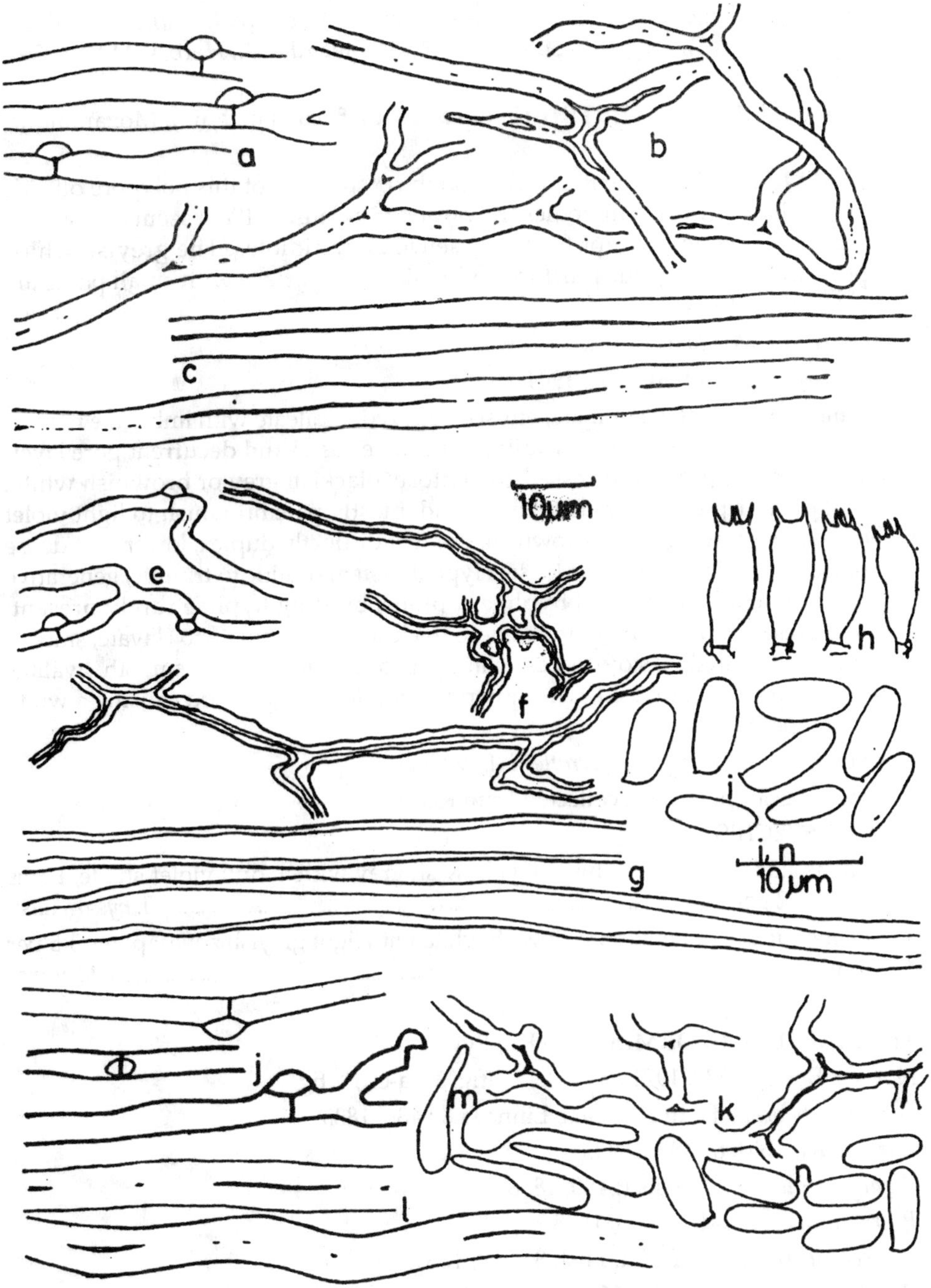

Fig. 30 : a-c *Trametes modesta* : a - generative hyphae; b - binding hyphae; c - skeletal hyphae. e — i *Trametes scabrosa* : e - generative hyphae; f - binding hyphae; g - skeletal hyphae; h - basidia; i - basidiospores.

j — n *Trametes varians* : j - generative hyphae; k - binding hyphae; l - skeletal hyphae; m - cystidioles; n - basidiospores.

SPECIMEN EXAMINED : On cut branches of *Terminalia paniculata* Roth. (COMBRETACEAE), Kuppadi, Sultan Battery, Wyanaad dist., Herb. CALI No. 294, 20-6-1986.

GEOGRAPHICAL DISTRIBUTION : East Africa (Tanzania), Mozambique, South Africa, India.

REMARKS : This is the first record on the occurrence of this polypore outside Africa. The East African fruitbodies (Ryvarden & Jonansen, 1980) seem to resemble the present collection except for the absence of cystidioles. The greyish white, scrupose and azonate pilues surface and white pore surface with small pores are characteristic of the species.

TRICHAPTUM Murr.

Bull. Torrey Bot. Club 31 : 608, 1904.

Fruitbody annual, resupinate, effused-reflexed to pileate with imbricate narrow pilei or with an elongated narrow pileus over an effused and decurrent pore layer. Pileus surface hispid to adpressed-tomentose, blackish grey or brownish white. Hymenophore irpicoid, lamellate, or poroid, mostly pale brownish to light violet when actively growing, tubes brownish. Context distinctly duplex, lower part dense and dark, upper part white and loose. Hyphal system dimitic to trimitic, generative hyphae with clamps; skeletal hyphae copious; binding hyphae rarely present. Cystidia present in the hymenium, thin- to thick-walled, subulate to clavate, smooth or apically encrusted. Basidiospores cylindrical, often slightly bent, smooth, hyaline, thin-walled and nonamyloid. On coniferous and deciduous wood causing a white rot. Cosmopolitan.

The species. *Trichaptum perrottetti* (Lev.) Ryv.

Two species heve been collected from Kerala.

KEY TO SPECIES

1. Pileus surface villose to hispid, yellowish grey with a dull violet shade; Pores large, 1-2 per mm.......... *T. byssogenus*

1′. Pileus surface tomentose, yellowish white with light greyish zones; pores 3-5 per mm.......... *T. biformis*

TRICHAPTUM BIFORMIS (Fr. in Kl.) Ryv.

Norw. J. Bot. 19 : 237, 1971. fig. 31. a-c., Pl. fig.

Polyporus biformis Fr. In Klotzsch, Linnaea 8 : 486, 1833.

P. friesii Kl., Ibid. 8 : 487, 1833.

P. pargamenus Fr. Epicrisis P. 480, 1838.

P. prolificans Fr., Ibid. p. 443, 1838.

P. laceratus Berk., Ann. Nat. Hist. 3 : 392, 1839.

P. xalapensis Berk. & Curt., Hook. Lomd. J . Bot. 1 : 103, 1849.

Polystictus pargamenus Fr., Nova Acta 3, 1 : 85, 1851.

Coriolus pargamenus (Fr.) Murr., North Am. Fl . 9 : 27, 1907.

C. prolificans (Fr.) Murr., North, Mycol. Notes 66 : 1192, 1922.

P. albovestidus Lloyd, Ibid. 69 : 1192, 1923.

Hirschioporus pargamenus (Fr.) Bond. & Sing., Ann. Mycol. 39 : 63, 1941.

Coriolus albovestidus (Lloyd) G. H. Cunn., Proc. Linn. Soc. N. S. W. 75 : 216, 1950.

Trichaptum pargamenum (Fr.) G. H. Cunn., N. Z. Dep. Sci. Industr. Res. Bull. 164 : 100, 1965.

Fruitbody annual, sub-stipitate to sessile, lateral with a narrow converging base, imbricate and confluent, flabelliform to dimidiate, often deeply lobate towards margin, thin, coriaceous while fresh, rigid when dry; 2.2-6.7 × 1.8-4.5 × 0.1-0.3 cm. Pileus surface pale yellow (3A3) to yellowish white (3A2) or wine yellow (3B3) to cream (4A3) with concentric zones of nougat (5D3) and its lighter shades, darker towards base; velvety in appearance, matted tomentose, abhymenial hairs protruding to form a cottony mass and more concentrated towards base; margin thin, often membraneous, entire or rarely splitting, rigid while fresh and dry. Pore surface yellowish white (4A2) to orange white (5A2) or white with a yellowish shade becoming butter yellow (4A5) when dry, yellowish white (3A2) towards margin, young fruitbodies pinkish white (11A2); pores absent near margin, visible to naked eye, oval to round, dissepiments projecting irregularly to appear irpicoid or daedaleoid, sometimes truely daedaleoid, 3-4 (5) per mm; pore mouth 120-160 μm broad; dissepiments 60-100 μm thick; pore tubes concolorous with the pore surace, up to 1 mm long, sunk into almost equal depths. Context cream (4A3), sometimes slightly greyish, homogeneous, slightly fibrous, up to 2 mm thick.

Hyphal system dimitic, generative hyphae hyaline, thin to slightly thick-walled, branched, septate and clamped, 2.5-3 μm in diam. Skeletal hyphae hyaline to pale-coloured, thick-walled with obliterate lumen or solid, unbranched and non-septate, cylindrical and straight, 3.5-7.5 μm in diam. Cystidia numerous; hyaline, cylindrical to clavate or ventricose with a blunt apex, slightly thick-walled, capitate, encrusted, lamprocystidia of muricate type, 15-22 × 4–5.5 μm. Basidia clavate, 4-spored, 12-13.5 × 3-3.75 μm. Basidiospores not observed.

DECAY : White rot; white mycelial growth in wood crevices.

SPECIMEN EXAMINED : On fallen branches, host unidentified, Nilambur Forest Division, Malappuram, Dist., Herb. CALI No, G142, 6-1-1985; Silent Valley National Park, Palghat Dist., Herb. CALI NO. G228, 15-9-1985; Chembra Peak, Meppadi, Wyanaad Dist., Herb. CALI No. G237, 11-9-1985.

GEOGRAPHICAL DISTRIBUTION : Generally a temperate taxon; North America, Europe, Finnoscandia, Australia, New Zealand, India, East Africa.

REMARKS : Bakshi's report (1971) of this fungus as *Polyporus pargamenus* shows narrower skeletal hyphae and shorter cystidia. The violet colour of the fresh pore surface exhibited by Finnoscandean (Ryvarden, 1978) and East African (Ryvarden & Johansen, 1980) sporophores were observed here only with young fruitbodies. The pale yellow fruitbodies with tomentose pileus surface, pale violet to yellowish white pore surface, irpicoid to daedaleoid dissepiments and capitate encrusted cystidia are charactertistic of this species.

TRICHAPTUM BYSSOGENUS (Jungh.) Ryv.

Norw. J. Bot. 19 : 372, 1972. fig. 31. d-g., Pl. fig.

Polyporus byssogenus Jungh., Verh. Batav. Genootsch. 17 : 43, 1838.

Trametes versatilis Berk., Hook. Lond. J. Bot. 1 : 150, 1842.

Polyporus venustus Berk., Ibid. 4 : 55, 1845.

P. cladotrix Berk. & Curt., J. Linn. Soc. 10 : 309, 1868.

Polystictus versatilis (Berk.) Fr. Nova Acta 3, 1 : 92, 1851.

Polyporus versatilis (Berk.) Rom., K. Svenska Vetensk akad. Handl. 26, 3(16) : 35, 1901.

Trichaptum versatile (Berk.) G. H. Cunn., N. Z. Dep. Sci. Industr. Res. Bull. 164 : 99, 1965.

Fruitbody annual, pileate, sessile, attached with a broad lateral base, imbricate and confluent, semi-circular to dimidiate, sometimes irregular to conchate in appearance, light corky in consistency, 1.5-4.5 × 1-3 × 0.2-0.3 cm. Pileus surface yellowish grey (4B2) with dull violet (14D3) hairs, colour of the hairs in concentric bands, adpressed with age; margin cerrulate, thin, rigid. Pore surface almost greyish Magenta (14E3) while fresh, dark blonde (5D4) when dry; pores irpicoid to somewhat daealeoid, poroid near margin, angular, up to 1.5 per mm; dissepimennts 165-200 µm thick; pore tubes slightly greyish orange (5B3), 1-2 mm long, sunk into equal depths, straight. Context concolorous with the pore tubes, duples, with a dense lower layer and loose upper layer, less than 1 mm thick.

Hyphal system dimitic. Generative hyphae hyaline, slightly thick-walled with broad lumen, branched, septate with clamps, common in dissepiments, 2-3 µm in diam. Skeletal hyphae hyaline to slighty yellowish, thick-walled to solid, usually unbranched, non-septate, cylindrical and equal in width, up to 4 µm in diam. Cystidia abundant, sub-hyaline, ventricose to ventricose-rostrate, lamprocystdiate, apically encrusted, thick-walled, thickness more pronounced towards the apex, 15-21 × 4.5-6 µm. Basidia clavate to broadly clavate, mostly bispored, with rudimentary sterigmata, 12-15 × 4.5-6 µm. Basidiospores not observed.

DECAY : White pocket rot.

SPECIMEN EXAMINED : On decaying fallen trunks of *Xylia xylocarpa* (Rozb.) Theob. (MIMOSAE), Kamathalamudi, Teak Plantation Division, Tunacadavu, Parambiculam, Palghat Dist., Herb. CALI No. G216, 12-7-1985. G. 268 12-11-1985.

GEOGRAPHICAL DISTRIBUTION : North and south America, West Indies, Australia, New Zealand, India, Philippines, Java, Africa.

REMARKS : This fungus has been reported earlier from India (Bose, 1937a; Bakshi, 1971) as *Polyporus versatilis*, Lloyd's (1898-1925) description of the fungus from Philippines closely agree with the present one.

TYROMYCES Karst.

Rev. Mycol. 3, 9 : 17, 1881.

Fruitbody resupinate to pileate. Pileus surface finely tomentose to smooth, usually white, more rarely, reddish or brownish. Pore surface white to cream, rarely greyish brown or reddish when dry; pores regular, small to medium, round to angular. Context or subiculum white to light brown when dry, consistency soft when

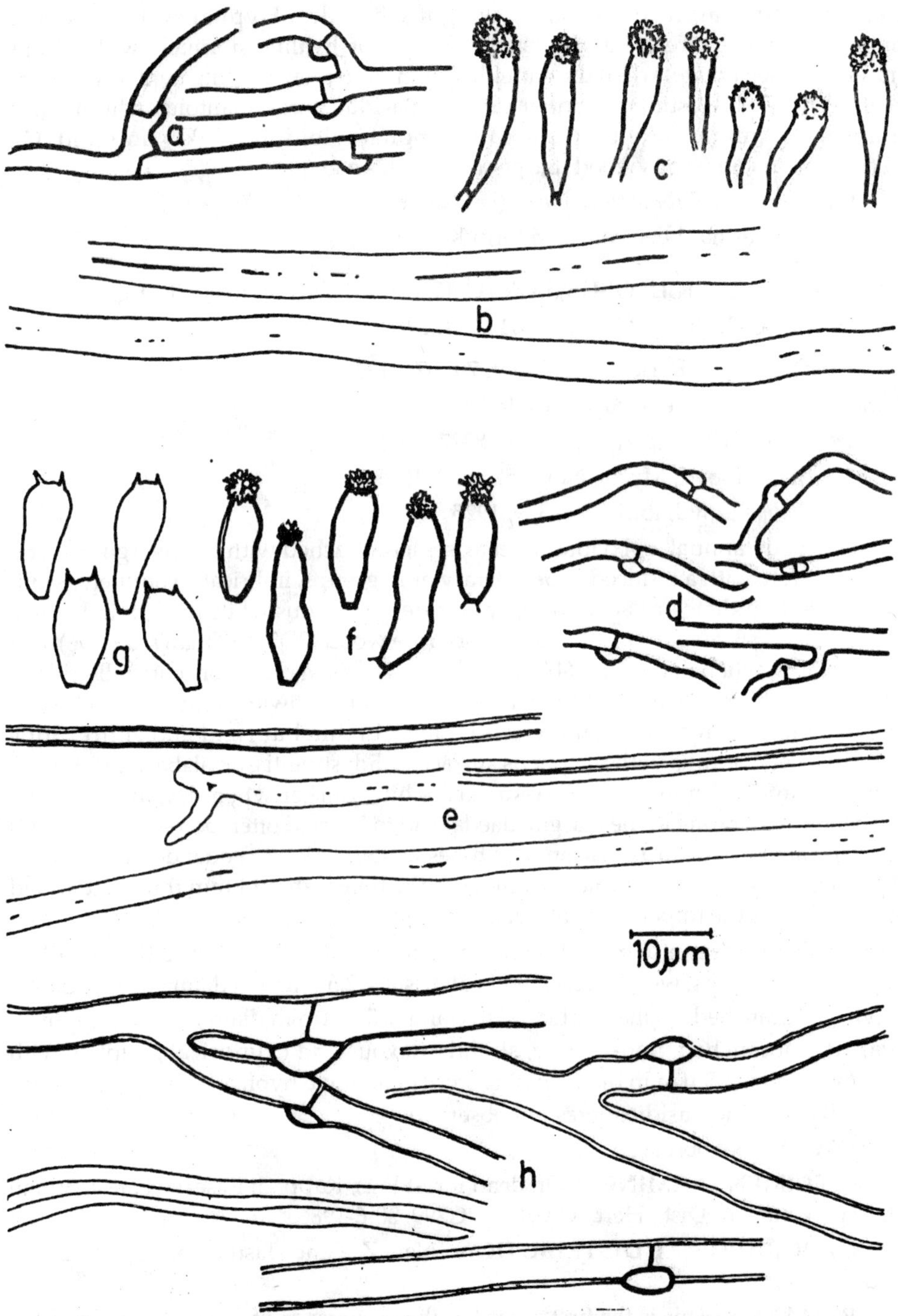

Fig. 31 : a-c *Trichanptum biformis* : a - generative hyphae; b - skeletal hyphae; c - cystidia. d - g *Trichaptum byssoqenus* : d - generative hyphae; e - skeletal hyphae; f - cystidia; g - basidia. h - *Tyromyces pelliculosus* : h - generative hyphae.

fresh, mostly fragile and light in weight when dry. Hyphal system usually monomitic, more rarely dimitic, generative hyphae hyaline, normally with clamps (rarely simple-septate) when dimitic either with rather few binding hyphae or hyline skeletal hyphae. Cystidia absent or present. Basidiospores allantoid, cylindrical to ellipsoid, hyaline, smooth, thin-walled, nonamyloid to weakly amyloid. On deciduous or coniferous wood. Large cosmopolitan genus.

Type species : *Tyromyces chioneus* (Fr.) Karst.

One species has been collected form Kerala.

TYROMYCES PELICULOSUS (Berk.) G. H. Cunn. fig. 30. h, Pl. , fig.
N. Z. Dept. Sci. Industr. Res. Bull. 164 : 124, 1965.

Polyporus pelliculous Berk., Lond. J. Bot. 7 L 575, 1848.

P. hystriculus Cke., Grevillea 15 : 16, 1886.

P. pelles Lloyd, Syn. Sect. Apus Polyp. P. 327. 1951.

P. atro-hispidus Lloyd, Mycol, Notes 57 : 823, 1919.

P. strigose-albus Lloyd, Ibid, 73 : 1329, 1925.

Fruitbody annual, epixylous, pileate, sessile, attached with a converging lateral base or rarely with a reduced stipe, solitary or in groups, imbricate, confluent, semi-circular to dimidiate or flabelliform, applanate, coriaceous while fresh, tough when dry; 2-5.5 × 2.-5 × 0.3-0.7 cm. Pileus surface rose wood (9D5) to Titian red (7D6), oak brown (5D6) when dry, cream (4A3) to white towards margin, concentrically ridged and zoned, hairs prominent but more concentrated towards the base to form a velvety mat, often strigose at the centre; margin thin but rigid and entire. Stipe reduced, concolorous with the pileus surface, solid; stipe-tissue alabaster (5B2), 1-2 mm long and 3-4 mm thick. Pore surface white, darker when dry, almost even, hymenium extending to the margin, daedaleoid to irpicoid often poroid, pores 1-2 (4) per mm, angular towards margin pore tubes white, sunk into equal depths, straight, 2.5-4 mm long. Context duplex, white near the tubes, about l mm thick. Titian red (7D6) towards the upper crust, about 2 mm thick.

Hyphal system monomitic. Generative hyphae hyaline, thin to thick-walled, septate with clamps, well represented in all tissues, 2.5-5 μm in diam.; upper context having unbranched hyphae, sptate with clamps, 5-6 μm in diam.; sometimes thick-walled, unbranched, non-septate, slightly flexous and cylindrical , hyphae with narrow lumen, 3-5 μm in diam. arising from generative hyphae. Cystidia and setae none. Basidia and basidiospores not observed.

DECAY : Not observed.

SPECIMEN EXAMINED : On dead forest logs, Kuppadi Reserve Forest, Sultan Battery, Wyanaad Dist., Herb. CALI No. G299, 20-6-1986.

GEOGRAPHICAL DISTRIBUTION : New Zealand, Tasmania, Central Africa, India.

REMARKS : This is ths first record of the occurrence of *T. pelliculosus* in India. The present collection agrees closely with the description of this polypore by Lloyd (1915b) and Ryvarden & Johansen (1980).

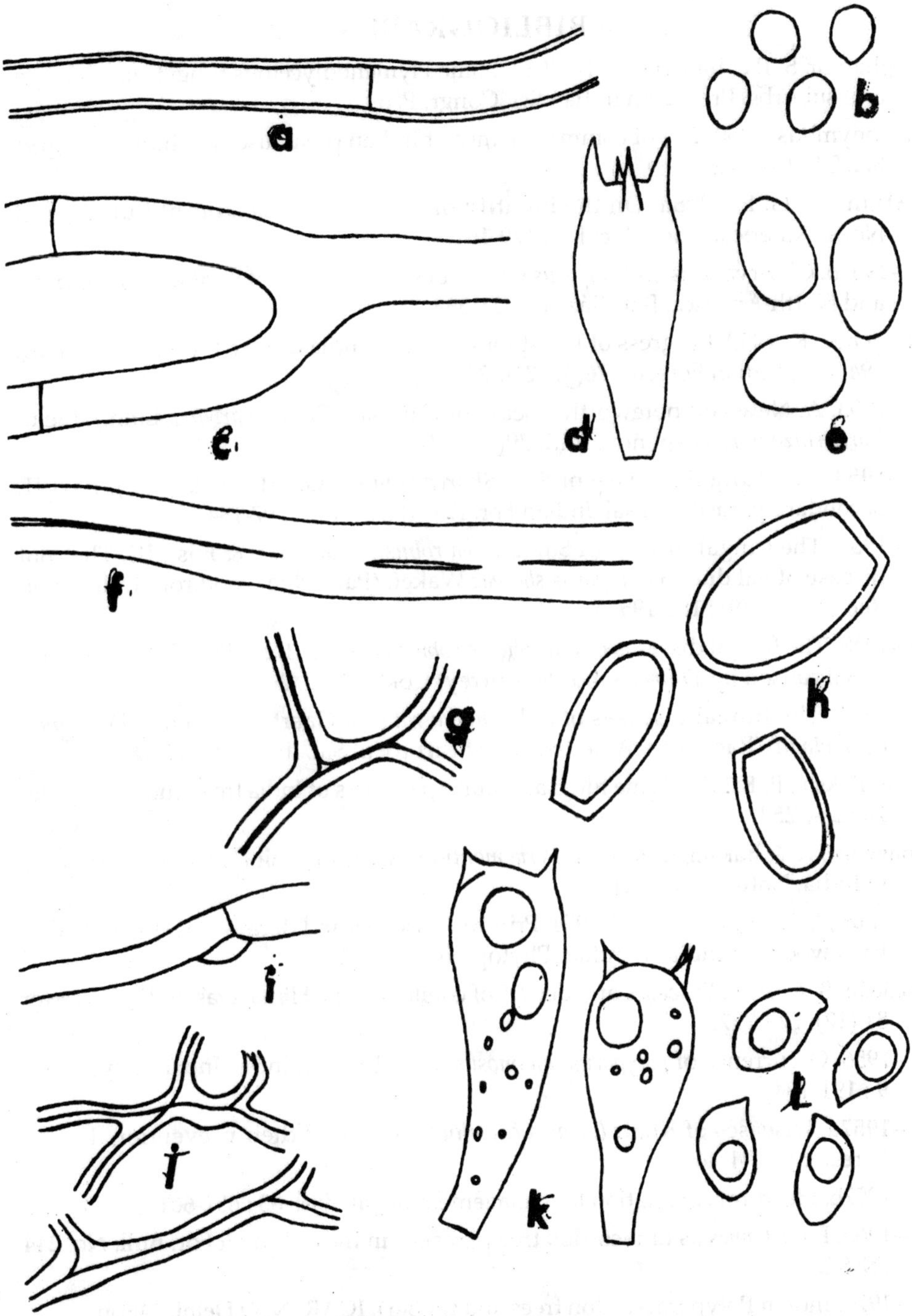

Fig. 32 : a & b. *Phylloporia chrysita* : a - generative hyphae; b - basidiospores.
c - e. *Coltrichia focicola* : c - generative hyphae; d - besidium; e - basidiospores.
f - h. *Perenniporia ochroleuca* : f - skeletal hyphae; g - binding hyphae; h - basidiospores.
i - l. *Polyporus meridionalis* : i -generative hyphae; j - skeletal hyphae; k-basidia; l -basidiospores.

BIBLIOGRAPHY

Agharkar, S. P. & Banerjee, S. N. 1933. Some Hymenomycetous fungi from Calcutta and suburbs. Proc. 20th Indian Sci. Congr. P. 69.

Anonymous. 1950. List of common names of Indian plant diseases. Indian J. Agric. Sci. 20 : 107-142.

Atkinson, G. F. 1908a. On the identify of *Polyporus applanatus* of Europe and North America. Ann. Mycol. 6 : 179-191.

—1908b. Observations on *Polyporus lucidus* Leyss. and some of its allies from Europe and North America. Bot. Gaz. 46. 321-338.

Bagchee, K. 1950. Progress of forest pathology in India during the quinquennium 1944- 49. Indian Forester 76(5) : 216-220.

—1953. A. New and noteworth disease of Gamhar (*Gmelina arborea* Linn.) due to *Poria rhizomorpha* sp. nov. Ibid. 79(1) : 17-24.

—1954. The fungal disease of Sal (*Shorea robusta* Gaertn. f. & Fils.), Part II, Secondary parasites of sal. Indian For. Rec. (N. S.) 1 (8) : 97-184.

—1957. The fungal disease of Sal (*Shorea robusta* Gaertn. f. & Fils), III. The root disease of Sal due to *Polyporus shoreae* Wakef. (Patrridge wood rot). Indian For. Rec. (N. S.) 1(9) : 185-195.

—1958. The fungal diseases of Sal (*Shorea robusta* Gaertn. f. & Fils), V. The heart rot of Sal caused by *Trametes incerta* (Currey) Cokke. Ibid. 2(4) : 61-69

—1961. The fungal diseases of Sal (*Shorea robusta* Gaertn. f. & Fils), IV. *Fomes caryophylii* (Rac.) Bres. A destructuve heart rot of Sal. Ibid. 2(3) : 25-28.

—& Bakshi, B. K.1950. Some fungi as wound parasities of India trees. Indian Forester 76 : 244-253.

Bagchee, K. & Bakshi, B. K. 1951. *Poria monticola* Murr. on Chir (*Pinus longifolia* Roxb) in India. Nature 167 : 824.

—, Puri, Y. N. & Bakshi, B. K. 1954. Principal diseases and decays of Oaks, and other hard woods in India-II. Indian Phytopath. 7 : 18-42.

Bakshi, B. K. 1955. Diseases and decays of conifers in the Himalayas. Indian Forester 81 (12): 779-797.

—1956. Occurrence of *polyporus squamosus* (Hund.) Fr. in India. Indian Phytopath. 9 : 191-194.

—1957a. Diseases of Khair (*Acacia catechu* Willd.) and their prevention. Indian Forester 83 : 41-46.

—1957b. Heart rots in relation to management of Sal. Ibid. 83 : 651-661.

—1965. Four *Fomes* as unrecorded tree parasites in India. Indian For. Bull. No. 244 (N. S.).

—1971. Indian Polyporaceae (on trees and timber). ICAR, New Delhi, 246pp.

—& Balwant, Singh. 1961. Heart rot and decay due to *Polyporus palustris*. Indian Forester 87 : 116-118.

—& Boyce, J. S. 1959. *Polyporus shorea* (root rot) on Sal. Ibid. 85 : 656-658.

Bakshi, B.K. & Sujan Singh. 1970. Heart rot in trees. Int. Rer. For. Res. 3 : 197-251.

—, Sen Manjusri & Singh, B. 1970. Cultural diagnosis of Indian Polyporaceae- II. Genera *Fomes* and *Trametes.* Indian For. Rec. (N. C.) 2(10) : 245-476.

—, Sehgal, H.S. & Singh, B. 1969. Cultural diagnosis of Indian Polyporaceae- I. Genus *Polyporus.* Indian For. Rec. (N. S.) 2(9) : 205-244.

Bakshi, B. K., Sujan Singh & Balwant Singh. 1963. A re-examination of *Fomes lignosus* and *Polyporus zonals.* Trans. Brit. Mycol. Soc. 46 : 426-430.

—& Ujagar Singh. 1966. A new root rot disease complex in teak. Indian Forester 92 : 566-569.

Balwant Singh. 1961. Occurrence of *Trametes ravida* (Fr.) Pilat in India. Ibid. 87 : 429-430.

—1966a. Timber decay due to five species of *Fomes* as new records. Ibid. 92 : 653-655.

—1966b. Studies on Indian *Poria II.* Diagnosis of five species as new record Ibid. 92 : 680-683.

— & Bakshi, B. K. 1961. New records of *Fomes* from India. Ibid . 87 : 302-303.

Banerjee, S. N. 1946. Some higher fungi of Sikkim-Himalayas. Sci. & Cult. 11 : 444-445.

—1947. Fungus Flora of Calcutta and suburbs I. Bull. Bot. Soc. Bengal 1 : 37-54.

—& Bakshi, B.K. 1945. Studies in the Biology of wood rooting fungi of Bengal. J. Indian Bot. Soc. 24 : 73-93.

—& Chakravarthy, M. 1945. On the biology of *Polyporus agaricus* Berk. Proc. 32nd Indian Sci Congr. p. 69.

—& Chatterjee, S. B. 1945a. Some observations on the identity of *Amauroderma rugosum* Nees., a rare species in Bengal. Proc. 32nd Indian Sci Congr. p. 68.

—&—1945 b. A note on the hyphal systems of the sporophores of *Amauroderma rugosum* Nees. Ibid. p. 68.

—&—1945c. Studies on "white rot" and "brown rot". Ibid. p. 68.

Banerjee S. N., & Ghosh, T. 1942-43. Preliminary report on the occurrence of higher fungi on bamboos in and about Calcutta. Sci. & Cult. 8: 194.

—&—1945. On the collection of Hymenomycetous fungi from the Sikkim-Himalayas. Proc. 32nd Indian Sci. Congr. p. 69.

—& Sarkar Anjali. 1956. Formation of sporophores of *Ganderma lucidum* (Leyss.) Karst. and *G. Applanatum* (Pers.) Pat. in Culture . Indian J. Mycol. Res. 2 : 80-82.

—& Sinha, A. K. 1955. A simple method for producing typical sporophores of *Polystictus sanguineus* (L.) Mey. Sci. & Cult. 20 : 612-613.

Baxter, D. V. 1926. Some porias from the region of the lake states. Papers of the Michigan Acad. of Sci., Arts & Letters VI : 67-76.

Bazzalo, Maria & Wright, J. E. 1982. Survey of the Argentine species of the *Ganoderma lucidum* complex. Mycotaxon 16; 293-325.

Berkeley, M. J. 1839. Description of exotic fungi in the collection of Sir W. J. Hooker, from memoirs and notes of J. F. Klotzsch, with additions and corrections. Ann. Nat. Hist. 3 : 375-401.

Berkeley, M.J. 1847. Decades of fungi XV-XIX. Ceylon Fungi No. 183. Lond. J. Bot. 6 : 312-326.

—1850. Decades of fungi XXV-XXX. Sikkim-Himalayan fungi. Hook. J. Bot. 2 : 42-51.

—1851. Decades of fungi XXXI-XXXVI. Sikkim-Himalayan fungi collected by Dr. Hooker. Hook. J. Bot. 3 : 4-21, 39-49, 77-84, 167-172, 200-206.

—1854. Decades of fungi XLI-L. Indian fungi. Ibid. 6 : 129-143, 161-174, 204-212, 225-35.

Berkeley M. J. 1855. Fungi, in Hooker, W. J. (ed.) Botany of the Antartic Voyage II. Flora Nova-Zealandia 2. London.

—1866. Fungi of the plains of India. Intellectual Observer 12 : 18-21.

—1872. Australian fungi received principally from Baron F. Von Mueller and Dr. R. Schomburk. J. Linn. Soc. Bot. 13 : 155-177.

Blatter, E. 1911. A list of Indian fungi, chiefly of the Bombay Presidency, with the description of two new species. J. Bombay Nat. Hist. Soc. 21 : 146-152.

Bondarstev, A. S. 1953. Polyporaceae of European Russia and the Caucasus. Moscow, Leningrad (in Russian).

Bose, S. R. 1919-1928. I. Description of fungi in Bengal. Proc. Indian Assoc. Cult. Sci. 4 : 119-114; II. Proc. Sci. Convention Indian Assoc. Cult. Sci. For the year 1918 : 136-143, III. Fungi of Bengal. Polyporaceae of Bengal Part III, Bull. Carmichael Med. Coll. Belgachia l : 1-5; IV. Polyporaceae of Bengal Part IV, Ibid. 2 : 1-5; V. Ibid. 3 : 20-25; VI. Part VI, Proc. Sci. Convention, Indian Assoc. Cult. Sci. for the year1919; VII. Ibid. For the year 1920-21; 27-36; VIII. J. Dept Sci. Calutta Univ. 9 : 27-34; IX. Ibid. 9 : 35-44.

—1921 a. One new species of Polyporaceae and some Polypores new to Bengal. Ann. Mycol. 19 : 129-131.

—1921b. Two new species of Polyporaceae. J. Indian Bot. 2 : 300-301.

—1922a.Une Polyporacee nouvelle du Bengale Bull. Soc Mycol. de France 38 : 173.

1922b. Geographical distribution of the Bengal species of Polyporaceae with a short history of them in Bengal. Proc. 9th Indian Sci. Congr. P. 117, also published in J., Indian Bot. Soc. 3 : 19-21.

Bose S. R. 1923 Une Polyporacee Nouvelle de l'Inde. Bull. Soc. Mycol. de France 39 : 226.

—1924. Les Polyporacees du Bengale. Revue Path. Veg. et Ent. Agric. 11 : 134-149.

—1925. A new species of Polyporaceae from Bengal. Ann. Mycol. 23 : 179-181.

—1929a. Revival of an old fruitbody of *Hexagonia discopda* Pat. & Har., and its successful spore culture from its fresh spore-discharge. Ibid. 27 : 321-323.

—1929b. Artifical culture of *Ganoderma lucidum* (Leyss) Fr. from spore to spore. Proc. 15th Indian Sci. Congr. P. 229.

—1931. A short note on the abnormal anatomy of the sporophore of *Ganoderma lucidum* (Leyss.) Fr. Proc. 18th Indian Sci. P. 263.

—1933. Abnormal spores of some *Ganoderma*. Mycologia 25 : 431-434.

Bose, S.R. 1934. Polyporaceae of Bengal X. J. Dep. Sci. Calcutta Univ. 11 : 1-18.

—1935. The distribution of polypores at high altitudes. Proc. 22nd Indian Sci. Congr. P. 251, also published in Ann. Mycol. 33. 201.

—1936. A note on the variation of spores of *Polystictus xanthopus* Fr. and *P. flabelliformis* Kl. at high altitudes. Tras. Brit. Mycol. Soc. 20 : 185.

—1937a. Polyporaceae from Lokra Hills (Assam). Ann Mycol. 35 : 119-137.

—1937b. A fringe within the pore-tubes of *Daedalea flavida* Lev. Proc. 24th Indian Sci. Congr. P. 259.

—1938. The presence of encrusted cystidia in the hymenium of *Polyporus zonalis.* Mycologia 30 : 683-684.

Bose S. R. 1939. An abnormal sterille form of *Polystictus sanguineus* Linn. Proc. 26th Indian Sci. Congr. P. 116.

—1942. Researches on Bengal polyporaceae. Univ. of Calcutta, Kirtikar Memorial 1 : 1-101.

—1944. Importance of anatomy in systematics of Polyporaceae,. J. Indian Bot. Soc. 23 : 153-157.

—1946. Polyporaceae of Bengal XI. J. Dep. Sci. Calcutta Univ. 2 : 53-87.

Bourdot, H. & Galzin, A. 1927. Hymenomycetes de France. 76lpp. Sceau 1972. (Reprint Lehre 1969).

Butler, E. J. 1903. A Deodar disease in Jaunsar. Indian. Forester 29 : 1-8.

—1909. *Fomes Lucidus* (Leyss.) Fr., a suspected parasite. Ibid. 35 : 514-518.

Butler, E. J. 1918. Fungi & Diseases in Plants. Thacker, Spink & Co. Calcutta, 547pp.

— & Bisby, G. R. 1931. The Fungi of India. Imp. Coun. of Agr. Res. India, Calcutta, Sci. Monogr. l, 237pp.

Cartwright, K. St. & Findlay, W. P. K. 1958. Decay of timber and its prevention. H. M. Stationery Office, London, 294pp.

Chandrasekharen, C. 1962a. A general note on the vegetation of Kerala State. Indian Forester 88 : 440-441.

—1962b. Forest types of Kerala - I. Ibid. 88 : 660-674.

—1962c. Forest types of Kerala - 2. Ibid. 88 : 731-747.

—1962d. Forest types of Kerala State - 3. lbid. 88 : 837-847.

Coleman, L. 1927. Structure of the spore wall in *Ganoderma.* Bot. Gaz. 83 : 48-59.

Cooke, M. C. 1876. Some Indian fungi. Grevillea 4 : 114-118.

—1881. Some Exotic fungi. Ibid. 9 : 97-101.

Cook, M.C. 1891a. Additions to *Daedalea.* Ibid. 19 : 91-93.

—1891b. *Trametes* and its allies. Ibid. 19 : 91-93.

—1891c. *Favolus* and *Laschia.* Ibid. 19 : 105.

Corner, E.J.H. 1932. A *Fomes* with two systems of hyphae. Trans. Brit. Mycol. Soc. 17 : 51-81.

Corner, E.J. H. 1950. Monograph of *Clavaria* and allied genera. Oxford Univ. Press, London. 740pp.

—1953. The construction of polypores I. Introduction. Phytomorphology 3 ; 152-167.

Cunningham, G. H. 1965. Polyporaceae of New Zealand. N. Z. Dep. Sci. Industr. Res. Bull. 164 : 1-304.

David, A. 1982. Etude Monographique du genre *Skeletocutis* (Polyporaceae). Naturaliste Can. (Rev. Ecol. Syst.) 109 : 235-272.

—& Rajchenberg Maria. 1985. The pore fungi from French Antilles and Guiana. Mycotaxon 22 : 285-325.

—, Bernard Dequarte & Jean Louis Fiasson. 1982. Two new *Phellinus* with globose, Cyanophilous spores. Ibid. 14 ; 160-174.

De, A. B. 1977. Infertility of *Polyporus grammocephalus* Berk. Curr. Sci. 46 : 58-59.

—1981. Taxonomy of *Polyporus osteriformis* in relation to its morphologhical and cultural characters. Can. J. Bot. 59 : 1297-1300.

De, A. B. & Roy Anjali, 1978. Typical sporophore production of *Polyporus tricholoma* Mont. in culture. Curr. Sci. 47 : 472-473.

—& Roy Anjali. 1980. Studies on Indian Polyporaceae-VII. Morphological and cultural characters of *Polyporus hirsutus* Wulf. ex. Fr. Indian J. Mycol. Res. 18 : 25-32.

—&—1981. Studies on Indian Polypores-IV. Morphological and cultural characters of *Polyporus grammocephalus*. Mycologia 73 : 150-156.

Donk, M. A. 1933. Revision der niederlandischen Homobasidiomycete-Aphyllophoraceeae II. Meded. Nederl. Mycol. Ver. 22 : 1-278.

Donk, M.A. 1948. Notes on Malaysian fungi. I. Bull. Bot. Gard. Buitenz. 3, 17 : 473-482

—1960. The generic names proposed for Polyporaceae. Persoonia 1 : 173-302.

—1964. A conspectus of the families of Aphyllophorales. Persoonia 3 : 199-324.

Fidalgo, O. & Fidalgo, M.E.P.K. 1966. Polyporaceae from Trinidad and Tobago-I. Mycologia 58 : 862-904.

—&—1967. Polyporaceae from Trinidad and Tobago-II. Ibid. 67 : 833-869.

Fries, E. M. 1825. Systema orbis vegetabilis, I. Plantae Homonemeae. 37pp.

—1874. Hymenomycetes Europaei. Unsaliae, 755pp.

Furtado, J. S. 1962. Structure of the spore of the Ganodermoideae Donk. Rickia 1 : 227-242.

Ganesh, P. N. & Leelavathy, K. M. 1986. New Records of *Phellinus* from India. Current Science 55(15) : 726-728.

Ganesh, P. N., Leelavathy, K. M. & Ryvarden, L. 1988. *Piloporia indica* nov. species. Trans. Brit. Mycol. Soc. 9(1) : 169-171.

Gilbertson, R. L. 1979. The genus *Phellinus* (Aphllophorales : Hymenochaetaceae) in Western North America. Mycotaxon 9 : 51-89.

Gilbertson, R. L. & Ryvarden, L. 1986. North American Polypores - Vol. l. *Abortiporus Lindtneria.* Fungiflora, Oslo, 1-433.

Hanif Khan, A. 1910. Root infection of *Trametes pini* (Brot.) Fr. Indian Forester 36 : 559-562.

Heim, R. 1962. L' organisation architecturale des spores de Ganodermes. Rev. Mycol. 27 : 199-212.

Hennings, P. 1900. Fungi Indiae Orientalis. Hedwigia 39 : 150-153.

—1901. Fungi Indiae Orientalis II, CI. W. Gollana. 1900 colecti. Ibid. 40 : 323-345.

Imazeki, R. & Toki, S. 1954. Higher fungi of Asakawa Experiment Forest. Bull. Govt. Exp. Sta. Tokyo 67 : 19-71.

Ireneia, M. 1980. Sete Especies de Polyporaceae Novas para Portugal. Biol. Sco. Brot. Ser. 2, 53 : 647-662.

—1983. O Genero *Phellinus* Quel. Em Portugal. Revista de Biologia 12 : 77-108.

Ireneia M. 1986 Studies on Aphyllophorales of Portugal. The genus *Ganoderma* Karst. Int. J. Mycol Lichenol 2 : 183-204.

Klan. Jaroslav & Libuse Kotilova, k. 1982. Macrofungi from the West Caucasus. Ceska. Mykol. 36 : 20-29.

Klotzsch, J. F. 1832. Mycologische Berichtingungen. Linnaea 7 : 193-204.

—1833. Fungi exoticie collectionibus britannorum. Ibid. 7 : 478-490.

Kornerup, A. & Wanscher, J. H. 1967. Methuen Handbook of Colour. Methuen & Co. Ltd., London, 243pp.

Kotlaba, F. 1977. Revision of the *Polyporus* (Polyporales) described as new by J. Venenovsky. Sbornik Narodniko Muzea V. Praze, Acta Musci Natiolas Prague 31B (1975), 182 : 1-56.

Kriesel, H. 1969. Grundzuge eines naturlichen systems der Pilze. J. Cramer, Jena, 245pp.

Leveille, J. H. 1844. Fungi, in voyage dans l' Inde par Victor Jacquemont pendant les annees 1829-1932, Tome 4 : 179.

—1846a. Descriptions des champignons de l' herbier du Museum de Paris. Ann.Sci Nat. 3-5 : 111-167.

—1846b. Fungi, P. 164-204 in Gaudichaud, M. (ed.) Voyage autour du Monde Sur la corvette La Bonite. Botanique, Vol. I.

Lloyd, C. G. 1898-1925. Mycological notes nos. 1-75 : 1-364 : Cincinnati, Ohio.

—1904-1919. Mycological letters 1-69, each separately paged, Cincinnati, Ohio.

Lloyd, C. G. 1910a. Synopsis of the genus *Hexagonia.* Mycol. Writ. 3 : 1-46.

—1910b. Synopsis of the section *Micropus. Tabacinus* and *Funalis* of the genus *Polystictus.* Ibid. 3 : 49-70.

—1912. Synopsis of the stipitate polyporoids. Ibid. 3 : 95-208.

—1915a. Synopsis of the genus *Fomes.* Ibid. 4 : 211-288.

—1915b. Synopsis of the section Apus of the genus *Polyporus.* Ibid. 4 : 291-392.

Locquin, M. 1957. Bull. Jard. Bot. Brux. 27 : 560-562. Quoted from Singer, 1975.

Lowe, J. L. 1934. The Polyporaceae of New York State. Bull. N. Y. State Coll. Foestry, Syracuse Univ. Tech. Publ. 41 : 1-142.

—1942. The Polyporaceac of New York State (except *Poria*). Ibid. No. 60.

—1952. Perennial polypores of North America *Fomes* with coloured spores. Mycologia 44 : 228-38.

—1957. The polyporaceae of North America. The Genus *Fomes*. Bull. State Univ. N. Y. Coll. Forestry Tech. Publ. 80 : 1-97.

—1958. The genus *Poria* in North America. Llodia 21(2) : 100-114.

—1963. A synopsis of *Poria* and similar fungi form the tropical regions of the world. Mycologia 55 : 453-86.

—1966. The polyporaceae of North America. The gunus *Poria*. Bull. State Univ. N. Y. Coll. Forestry Tech. Publ. 90 : 1-183.

Lowe, J. L. & Gilbertson, R. L. 1961. Synopsis of the polyporaceae of the Western United States and Canada. Mycologia, 53, 474-511.

Massee, G. 1901. Fungi Exotici 3. Kew Bull. 1901. : 150-169.

—1906. Fungi Exotici 4. Ibid. 1906 : 91-94.

—1908. Fungi Exotici 8. Kew Bull. 1908 : 216-219.

—1910. Fungi Exotici. Ibid. 1910 : 249-253.

Mitter, J. H. & Tandon, R. N. 1932. Fungus flora of Nainital, Part I, J. Indian Bot. Soc. 11 : 178-180.

—&—1938. Fungi of Nainital, Part II. Ibid. 17 : 177- 182.

Montagne, J. F. C. 1842. Cryptogamae Nilgherenses. Ann. Sci. Nat. 2, 18 : 12-23.

Murrill, W. A. 1905. The Polyporaceae of North America XI. A synopsis of the brown pileate species. Bull. Torrey Bot. Club 32 : 353-71.

—1915. Tropical polypores. (Reprinted) Bibliotheca Mycologica Band 40 (1973), 113 pp.

—1924. Kashmir fungi. Mycologia 16 : 133.

Natarajan, K. & Kolandavelu, K. 1985. Resupinate Aphyllophorales from South India-I. Kavaka 13(2) : 71-76.

Neuman, J.J. 1914. The Polyporaceae of Wisconsin. (Reprinted) Bibliotheca Mycologica. Band 32 (1971), 156pp.

Niemela, T. 1982. On Finnoscandian Polypores 8. New genus *Piloporia*. Karstenia 22: 13-16.

Niemela, T., & Uotila, P. 1977. Lignicolos macrofungi from Turkey and Iran. Karstenia 17 : 33-39.

Nobles, M.K. 1958. Cultural characters as a guide to the taxonomy and phylogeny of the Polyporaceae. Can. J. Bot. 36 : 883-926.

Overholts, L.O. 1953. The Polyporaceae of United States, Alaska and Canada. Univ. Mich. Stud. 19 : 1-466.

Overeem, C. van. 1925. Bull. Jard. Bot. Buitenz. 3, 7 : 434-436. Quoted from Donk, 1964.

—& Weese, J. 1924. Icones Fungorum Malayensium. 7. Vienna. Quoted from Singer, 1975.

Parmasto, E. 1983a. The spores of *Hexagonia apiaria* and *H. Tenuis* (Aphyllophorales). Mycotaxon 18 : 49-51.

—1983b. *Leucophellinus mollissimus,* a tropical polypore found in the Soviet Far East. Eesti NSV Teaduste Akadeemia Toimetised, 32. Koide Biologia. 1983, 4 : 264-272.

Patouillard, N. 1890. Sur la place du genre *Favolus* dons la classification . Bull. Sco. Mycol. Fr. 6 : 19-21.

—1900. Essai Taxonomique sur les Familles et les Genres des Hymenomycetes. Paris, 184 pp.

Peace, T.R. 1962. Pathology of trees and shrubs with special reference to Britain. Clarendon Press, Oxford, 753pp.

Pegler, D.N. 1973a. Poroid families of the Aphyllophorales. In Ainsworth, G.C., Sparrow, F.K. & Sussman, A.S. (ed.). The Fungi An advanced Treatise. Vol. IVB. 397-420.

Pegler, D.N. 1973b. The Polypores. Supplement Bull. Brit. Mycol. Soc. 7, 1 : 1-43.

— & Waterston, J.M. 1968. CMI descriptions of Pathogenic Fungi and Bacteria. Nos. 191-200.

Perreau-Bertrand, J. 1967. Researches sur la diferenciation et la structure de la paroi sporale chez les Homobasisiomycetes. A Spores ornees. Ann. Sci. Nat. Bot. Paris 12, 8 : 639-746.

—1973. Contribution a l'etude des ornaments sporaux chez les Ganodermes. Rev. Mycologie 37 : 241-252.

Persoon, C.H. 1801. Systema Methodica Fungorum. 706pp.

Peterson, J.E. 1987. *Ganoderma* in Northern Europe. Bull. Brit. Mycol. Soc. 21(2) : 62-67.

Puri, Y.N. 1956. Studies on Indian *Poria.* J. Indian Bot. Soc. 35 : 277-283.

Rajchebverg. M. 1982. El genero *Coriolus* (Polyporaceae) En la Republica Argentina. Bol. Soc. Argentina Bot. 21(1-4) : 17-57.

—1984. Basidiomicetos xilofilos de la Region Mesapotamica, Republica Argentina. V. Poliporos Resupinados. Revista de Investigaciones Agropecuarias INTA, Buenos Aires, Republica Argentina (19(1) : 1-105.

Rangaswami, G., Seshadri, V.S. & Lucy Channamma, K.A. 1970 Fungi of South India. Univ. Agric. Sci. Bangalore, 193pp.

Rea, C. 1922. British Basidiomycetes. Cambridge, 799pp.

Reid, D.A., Thind, K.S. & Chatrath, M.S. 1959. The Polyporaceae of Mussoori Hills, India-IV. Trans. Brit. Mycol. Soc. 42 : 40-44.

Roy Anjali. 1968a. Anatomy of India Polyporaceae-I. *Trametes cingulata* Berk. and *T. Persooni* Fr. Bull. Bot. Soc. Bengal 22 : 45-54.

Roy Anjali. 1968b. Anatomy of India Polyporaceae- II. *Daedalea flavida* Lev. and *D. Microzona* Lev. Ibid. 22 : 131-134.

—1969. Anatomy of Indian Polyporaceae-III. *Polyporus adustus* Willd. ex Fr. and *P. osteriformis* Berk. Ibid. 23 : 205-211.

—1971. Anatomy of India Polyporaceae- V. *Polyporus anthelminticus* Berk. Visva Bharati Annals (Sc.) Part II, 14 : 20-29.

—1972. Some microstructures in relation to Polyporaceae. Mycopath. Mycol. Appl. 48 : 111-119.

—1975. Anatomy of Indian Polyporaceae- VI. *Hexagonia discopoda* and *H. Sulcata.* Bull. Bot. Soc. Bengal 29 : 57-64.

—1976. Structures of zones in fruiting bodies of Polyporaceae. Nova Hedwigia 27 : 801-804.

—1981a. Studies on Indian polypores- VI. Morphological and cultural characters of *Irpex flavus* Klotzsch. Ibid. 34 : 259-263.

—1981b. Studies on Indian polypores- VIII. Morphological and cultural characters of *Ganoderma colossum* (Fr.) Torrend. Ibid. 35 : 749-754.

—& De., A. B. 1977. A record of *Polyporus trocholoma* Mont. From India. Trans. Brit. Mycol. Soc. 68 : 442-444.

—&—1979. Studies on Indian polypores- I. Morphological and cultural characters of *Polyporus anthelminticus.* Bull. Bot. Soc. Bengal 33(1/2) : 105-114.

Roy Anjali & De, A. B. 1980. Studies on Indian polypores — III. Morphological and cultural characters of *Trametes floccosa.* Norw. J. Bot. 27; 297-300.

—&—1996. Polyporaceae of India. Int. Book Dist. Dehra Dun, 309pp.

Roy, S. K. 1953. Notes on some Hymenomycetes new to India. Sci. &. Cult. 19 : 94-96.

Ryvarden, L. 1973. Type studies in the Polyporaceae - I. Tropical species described by C.H. Persoon. Persoonia 7(2) : 305-312.

—1976a. Polyporaceae of North Europe. Vol. I. Fungiflora, Oslo, 1-218.

—1976b. Type-studies in the Polyporaceae, 4. Species described by J. F. Klotzch. Mem. N.Y. Bot. Gard. 28(1); 199-207.

—1976c. Type-studies in the Polyporaceae, 7. Species described by M. J. Berkeley from 1836 to 1843. Kew Bull. 31(1) : 81-103.

—1976d. Type-studies in the Polyporaceae - 8. Species described by E. Rostrup. Bot. Tidsskrift bd. 71 : 100-102.

—1977. Type-studies in the Polyporaceae - 10. Species described by J.M. Berkeley, either alone or with other authors, from 1844 to 1855. Norw. J. Bot. 24 : 213-220.

—1978. The Polyporaceae of North Europe - Vol. 2. Fungiflora, Oslo, 219-507.

—1981. Type studies in the Polyporaceae - 13. Species described by J.H. Leveille. Mycotaxon 23 : 175-186.

Ryvarden, L. 1984. Type studies in the Polyporaceae- 16. Species described by J. M. Berkeley, either alone or with other mycologists from 1856 to 1886. Ibid. 20 ; 329-363.

—1986. *Piloporia sajanensis* (Parm.) Niemela, a rare polypore in Finnoscandia. Windahia 16 : 97-98.

—& Jonansen, I. 1980. A preliminary polypore flora of E. Africa. Fungiflora, Oslo, Norway, 636pp.

Sarkar, Anjali. 1969. Developmental anatomy and differentiation of tissue systems of *Ganoderma lucidum* (Leyss. ex Fr.) Karst. Phyton 13(20) 89-104.

Sehgal, P. M., Mukerjee, S. K. & Singh, B. 1961. Short note on the fungus flora of Nicobar Islands. Indian Forester 87 : 766-767.

—, Sen, M. & Bakshi, B. K. 1966. Temperature relationships of Indian polypores. Indian For. Rec. (N. S.) 2(7) : 131-137.

Sen Menjusri 1973. Cultural diagnosis of Indian Polyporaceae- III. Genera *Daedalea, Favolus, Ganoderma, Hexagonia, Irpex, Lenzites, Merulius* and *Poria*. Indian For. Rec.(N. S.) 2(11) : 277-304.

Sharma, J. K., Mohanan, C. & Maria Florence, E. J. 1985. Disease survey in nurseries and plantations of forest tree species grown in Kerala. Kerala Forest Research Institute, Research Report, 36 ; 1-268.

Sher Singh. 1924. The liability of Deodar to the attack of *Trametes pini* (Brot.) Fr. in Lolab, Kashmir. Indian Forester 50 : 361-365.

Singer, R. 1962. Agaricales in Moderm Taxonomy 2nd Ed. J. Cramer, Germany, 915pp.

—1975. Agaricales in Modern Taxonomy 3rd Ed. Cramer, Vadmer, Vaduz 912pp.

—1986. Agaricales in Modern Taxonomy 4th Ed. Sven Koeltz, Germany, 981pp.

Steyaert, R.L. 1967. Les *Ganoderma* palmicolas. Bul. Jard. Bot. Nat. Belg. 37 : 465-492.

—1972. Species of *Ganoderma* and related genera mainly of the Boger and Leiden herbaria. Persoonia 7 : 55-118.

—1975. The concept of circumscription of *Ganoderma tornatum*. Trans. Brit. Mycol. Soc. 65 : 451-467.

Sujan Singh, Singh, B.& Baakshi, B.K. 1961. Fungus flora of South Andamans. Indian Forester, 87 : 248-250.

Teixeira, A.R. 1962. The taxonomy of the Polyporaceae. Biol. Rev. 37 : 51-81.

Theissen, F. 1911. Fungi aliquot Bombayenses Rev. Ed. Blatter Collecti. Ann. Mycol. 9 : 153-159.

Thind, K.S. & Chatrath, M.S. 1957. Polyporaceae of Mussoorie Hills. II. Res. Bull. Punjab. Univ. 25 : 431-442.

—&—1960. Polyporaceae of Mussoorie Hills I. Indian Phytopath. 23 : 76-89.

—& Dhanda, R.S. 1978. The Polporaceae of India - XL. Ibid. 31 : 463-472.

—&—1979a. The Polyporaceae of India - IX. Eight species of *Poria* new to India. Kavaka 7 : 51-58.

Thind, K.S. & Dhanda, R.S. 1979b. The Polyporacea of India - XII. The genus *Albatrellus*. Indian Phytopath. 32 : 55-60.

—& Dhanda, R.S. 1980. The Polyporaceae of India - X. Kavaka 8 : 59-67.

—& Rattan, S.S. 1971a. The Polyporaceae of India - V. Indian Phytopath. 24 : 50-57.

—&—1971b. The Polyporaceae of India -VII. Ibid. 24 : 290-294.

Thind, K.S. & Rattan, S.S. 1971c. The Polyporaceae of India - VIII. Res. Bull. Punjab Univ. 22 : 27-34.

—, Bindra, P.S. & Chatrath, M.S. 1957. The Polyporaceae of Musssorie Hills - III.. Ibid. 12 : 471-483.

—, Rattan, S.S. & Dhanda, R.S. 1970. The Polyporaceae of India - VI. Ibid. 21 : 109--17.

Uppal, B.N., Patel, M.K. & Kamat, M.N. 1935. The fungi of Bombay - VIII. 1-56.

Vahid., S.A. 1938. Damage of *Acacia arabica* by *Fomes pappinus* Bres. Indian Forester 54 : 662-64.

Vijayan, K.M., Natarjan, S. & Krishna Murthy, C.S. 1973. Cocount wilt disease in Tamil Nadu Madras Agric. J. 60 : 504-506.

Wright, J.E. & Deschamps, J.R. 1972. Basidiomycetes xilofagos de los Bosques Andinopatagionicos. Revista de investigaciones Agropecuarias INTA Buenos Aires, Series 5. Patalogia Vegetal. 9(3) : 111-195.

INDEX OF GENERA AND SPECIES

HOST-INDEX

39. *S. mahogani* (Linn.) Jacq. 38
40. *Syzygium cumini* (Linn.) Skeels 22
41. *Syzygium* sp. 55
42. *Tectona grandis* Linn. F. 44, 62, 68, 72, 89, 102, 106, 125, 131, 143
43. *Terminalia alata* Heyne ex Roth. 34
44. *T. arjuna* (D. C.) Roxb. ex.-Wight & Arn. 36
45. *Terminalia bellerica* (Gaertn.) Roxb. 44, 122
46. *T. crenulata* Roth. 50
47. *T. paniculata* Roth. 37, 75, 134, 146
48. *Terminalia* sp. 45, 99
49. *Theobroma cacao* Linn. 24
50. *Toona ciliata* Roem. 22, 44, 101
51. *Vateria indica* Linn. 36, 40, 44
52. *Xylia xylocarpa* (Roxb.) Theob. 38, 148
53. *Zanthaxylum rhetsa* (Roxb.) D. C. 137.
